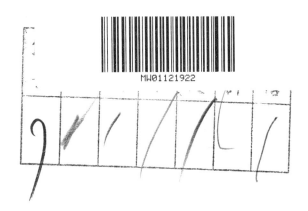

Were We the Enemy?

Transitions: Asia and Asian America
Series Editor, *Mark Selden*

Were We the Enemy? American Survivors of Hiroshima,
Rinjiro Sodei

North Koreans in Japan: Language, Ideology, and Identity,
Sonia Ryang

Proletarian Power: Shanghai in the Cultural Revolution,
Elizabeth J. Perry and Li Xun

*City-States in the Global Economy:
Industrial Restructuring in Hong Kong and Singapore,*
Stephen Wing-Kai Chiu, Kong-Chong Ho, and Tai-lok Lui

*The Taiwan-China Connection: Democracy and Development
Across the Taiwan Straits,* Tse-Kang Leng

Trade and Transformation in Korea, 1876–1945,
Dennis L. McNamara

Hidden Horrors: Japanese War Crimes in World War II,
Yuki Tanaka

*Encountering Macau: A Portuguese City-State on
the Periphery of China, 1557–1999,* Geoffrey C. Gunn

How the Farmers Changed China: Power of the People,
Kate Xiao Zhou

From Plan to Market: The Economic Transition in Vietnam,
Adam Fforde and Stefan de Vylder

A "New Woman" of Japan: A Political History of Katô Shidzue,
Helen M. Hopper

*Japanese Colonialism in Taiwan: Land Tenure, Development,
and Dependency, 1895–1945,* Chih-ming Ka

Vietnam's Rural Transformation,
edited by Benedict J. Tria Kerkvliet and Doug J. Porter

*The Origins of the Great Leap Forward:
The Case of One Chinese Province,* Jean-Luc Domenach

*The Politics of Democratization: Generalizing
East Asian Experiences,* edited by Edward Friedman

Our Land Was a Forest: An Ainu Memoir, Kayano Shigeru

Were We the Enemy?

American Survivors of Hiroshima

Rinjiro Sodei

Edited by
John Junkerman

A Division of HarperCollinsPublishers

Transitions: Asia and Asian America

All rights reserved. Printed in the United States of America. No part of this publication may be reproduced or transmitted in any form or by any means, electronic or mechanical, including photocopy, recording, or any information storage and retrieval system, without permission in writing from the publisher.

Copyright © 1998 by Westview Press, A Division of HarperCollins, Inc.

Published in 1998 in the United States of America by Westview Press, 5500 Central Avenue, Boulder, Colorado 80301-2877, and in the United Kingdom by Westview Press, 12 Hid's Copse Road, Cumnor Hill, Oxford OX2 9JJ

Library of Congress Cataloging-in-Publication Data
Sodei, Rinjirō, 1932–
    [Watakushitachi wa teki datta no ka. English]
    Were we the enemy? : American survivors of Hiroshima / Rinjiro Sodei ; edited by John Junkerman.
        p.   cm. — (Transitions—Asia and Asian America ; 204)
    ISBN 0-8133-2960-4 (hc)
    1. Hiroshima-shi (Japan)—History—Bombardment, 1945—Personal narratives.   2. World War, 1939–1945—Japanese Americans.
3. Americans—Japan.   4. Japanese—United States.   I. Junkerman, John.   II. Title.   III. Series.
D767.25.H6S6213   1998
940.54'26—dc21   97-45583
                                                                                                            CIP

The paper used in this publication meets the requirements of the American National Standard for Permanence of Paper for Printed Library Materials Z39.48-1984.

10   9   8   7   6   5   4   3   2   1

Contents

Foreword, John W. Dower	ix
Introduction	1
1 From Hiroshima, Back to Hiroshima	5
2 Death—and Life—in the Desert	20
3 Hiroshima: The Target City	28
4 Heading Toward the Ruined City	36
5 Nisei Coming, Nisei Going Home	48
6 Strangers in Their Own Homeland	57
7 Pieces of the Jigsaw Puzzle	70
8 The Death of the "President's Patient"	80
9 The Hibakusha Begin to Organize	88
10 Hibakusha Discovered	97
11 You Were Our Enemies!	109
12 In Search of Hibakusha	122
13 The Many Shades of the Hibakusha Experience	133
14 Ups and Downs	143
15 A Medical Team Comes and Goes	154
16 Washington Comes to Los Angeles	164
17 "We Are All Hibakusha"	173
18 Epilogue: Fifty Years After the Bomb	181

Foreword

It is always sobering to learn how long it takes to see things clearly. Even cataclysmic events like the atomic bombings of Hiroshima and Nagasaki, which almost mesmerized the entire world when they took place over a half century ago, have revealed their secrets, their full dimensions, only slowly.[1]

Take, for instance, what might seem to be a straightforward issue: the number of men, women, and children killed by the bombs. Early estimates by the U.S. government put these in the range of between 70,000 and 80,000 killed in Hiroshima and 35,000 in Nagasaki, and low figures of this order have survived in some sources to the present day. In fact, as became clear by the 1970s, the most credible calculations place fatalities at roughly double those numbers in each city. Around 210,000 individuals probably were killed by the two bombs (140,000 in Hiroshima and 70,000 in Nagasaki), most of these deaths occurring immediately or in the months that followed. To the present day, small numbers die each year of bomb-related effects; and to the present day, there are still no clearly disaggregated, cumulative, official figures on this seemingly most fundamental of questions.

Or take the question of what it is that defines a "hibakusha," or atomic-bomb victim. In Japan, formal definition did not come until 1957, when the government belatedly agreed to provide special free medical services to survivors. Under Japanese law, several categories of individuals qualify for such care: persons who were within four kilometers of the hypocenters of the bombs; those who entered the stricken cities within three days, and came within two kilometers of ground zero; and those who were in utero when their mothers became hibakusha. Because of the uncertain genetic effects of the nuclear irradiation, the Japanese government also extends medical care to children subsequently born to survivors.

With rare exceptions, persons who were within one kilometer of the hypocenters were killed instantly by some combination of blast, heat, fire, and radiation. To the Japanese and non-Japanese researchers who have studied the long-term medical effects of the bombs, by far the most vulnerable group of survivors was within approximately two kilometers of ground zero at the time of impact. How much radiation each individual may have been exposed to can only be estimated by a careful "radiation dosimetry" that takes into account the total amount of radiation released by each bomb, plus exactly where each survivor was, and how exposed or protected (by walls, terrain, or the like) he or she may have been. Before political pressure led the government to adopt broader criteria, Japanese physicians also focused on those who had been within this two-thousand meter radius when the bombs exploded.

As the imprecision of the concept of "survivor" suggests, it has been difficult to grasp the full dimensions of the atomic-bomb experience because that experience is still with us, still revealing new mutant faces. To speak of the *effects* of the bombs is to speak not merely of the early, gruesome mass deaths, but of pathologies that have exposed themselves only piecemeal—and with decidedly different impacts and meanings depending on whether it be to scientists, to detached laymen, or to the survivors themselves. Almost immediately, published photographs made the general public familiar with the hard, disfiguring "keloid" scars suffered by many burn victims. Soon afterwards, despite U.S. government attempts to repress such reportage, the horrendous "atomic-bomb disease" that cut through seemingly healthy survivors like a bloody scythe became well known: hair falling out in clumps, purple splotches on the skin, bloody vomit and diarrhea, thirst, fever, death.

This was but the start of the macabre process of learning what August 6 and August 9 had wrought. Within a few years, it became clear that the bombs had reached into the womb. Stillbirths and neonatal deaths were abnormally high among births to women who had been pregnant when exposed to the bombs. Among the surviving children, stunted growth, including small head size and a high incidence of severe mental retardation, was common. Beginning in 1947, researchers associated with the U.S.-sponsored Atomic Bomb Casualty Commission attempted to examine every newborn child in Hiroshima and Nagasaki (they succeeded in ninety-five percent of all households)—arriving in still shattered neighborhoods with great solemnity in their official vehicles, an occasion for all to behold. Where the child was stillborn or soon died, autopsies were requested—an unfamiliar practice in Japan, bordering on desecration (a majority of bereaved families nonetheless complied). Regular examinations of the children of survivors were carried out over the years that followed. Imagine being the parent in such a household, or the child, or a prospective spouse.

Soon afterwards, it was observed that the incidence of leukemia was higher than normal among youngsters who had been exposed to the bombs (the first scientific paper on this appeared in 1952). Leukemia preys on children; and children, particularly up to around the age of ten, are especially susceptible to adverse effects of irradiation. It made sense to anticipate more "adult" cancers, and these did indeed materialize. In time, it became known that survivors suffered a somewhat higher incidence of breast and thyroid cancer, among other maladies. The major scientific summation of these matters, based on research begun in 1950 and involving a sample of some 93,000 survivors (37,800 of whom had died), plus a control population, was not published until 1994; and its conclusions were *not* catastrophic. Apart from deformities in children who were in utero when the bombs were dropped, it is impossible to distinguish between cancer or non-cancerous diseases that occur naturally and pathologies that are radiation induced. In the 1994 study, deviance was in any case small: The incidence of cancer (some 8,000 deaths) among survivors was slightly more than 400 cases above the norm.

Survivors do not read scientific papers. What they know about is the indescribable shock they experienced, the atomic-bomb deaths and sicknesses they witnessed, the visits of investigators to mothers who had just given birth, the researchers who materialized at regular intervals in Japan to run batteries of tests not only on survivors but on their children, even those born afterwards. And the scientists, on their part, took an inordinately long time to grasp what then became rather lamely called the "human effects" of the bombs. What the scientists recognized belatedly—and non-Japanese have rarely thought about at all—is what in the Japanese turn of phrase is sometimes spoken of as the "keloids of the heart," the "leukemia of the soul." We grow old. If cursed, cancer or some such killer invades our bodies. But survivors have lived in daily imagination of such violation. For them, the trauma of the shattering experience itself was compounded by gnawing fear that the bomb still lived within the body.

So the presumably simple question of the "effects" of the bombs has also proved to be not simple at all. Like one of our present-day pharmaceutical capsules, its impact has been incremental, increasingly stronger. Much the same time-staggered effect also has characterized the manner in which we have slowly comprehended *whom* the bombs killed, maimed, and left as survivors. For the longest while, the answer here, at least, seemed obvious: the casualties were "Japanese." This has always been the position of those associated with dropping the bombs. And, out of a deep sense of victimization, many Japanese themselves have taken an almost perverse pride in depicting themselves as the sole sufferers in this terrible dawn of the nuclear age.

But this is myth. It took, again, several decades before it became widely acknowledged that as many as twenty-thousand conscripted Korean workers were in the two cities when the bombs were dropped, thousands of whom were killed instantly (Korea had been a Japanese colony since 1910). Victims of the two bombs also included small numbers of Chinese, Southeast Asians, and Europeans. One or two dozen Caucasian American prisoners of war were in Hiroshima when it was bombed. And—and this is among the last of the hidden facts to imprint on popular consciousness—probably around three thousand U.S. citizens of Japanese background were residing there when the city was destroyed. Hiroshima city and its environs, as Professor Sodei makes clear for us in this eye-opening book, was the most prolific source of Japanese emigrants to the United States. Many of their American-born children had returned there before the war, often as students, and were unable to return to their birthplace after the Japanese attacked Pearl Harbor.

The Hiroshima bomb killed over half of the people residing in the city and, by this ratio, took more than a thousand American lives by incinerating or irradiating Japanese Americans trapped there. The story told here in *Were We the Enemy?*, however, concerns those who survived the bomb. This is a painful story, in the "shadow of history," as Professor Sodei would put it. But it is also a story of Japanese American women and men—including, as it happened, female survivors of the bombs who married Americans and came to the United States as naturalized citizens—who found the courage to talk openly of their pain and stigmatization, and to demand their *rights* as American citizens. The individuals to whom Professor Sodei has given voice show us the keloids of the heart. And despite racial and gender discrimination, despite language handicaps and cultural conflicts, they tell us important things about America itself.

This is one of several unexpected outcomes of this narrative. The story Professor Sodei set out to document in the 1970s involved the ultimately successful struggle of American survivors in California to gain access to regular medical examinations—by visiting Japanese doctors—comparable to those provided hibakusha in Japan. The results of these exams, we learn through the author's follow-up coverage, have been essentially negative insofar as concerns pathologies that can be explicitly related to the atomic bombs. The few hundred American hibakusha who have participated in this program are a "statistically insignificant" group. They have shown no deviance from natural disease pathologies, and no undesirable genetic effects clearly traceable to the bombs—findings that, at long last, offer them a modicum of comfort.

At the same time, however, neither the American nor Japanese hibakusha can ever be "normal" as others understand this term, for their experience was, and remains, extraordinary. Although they would not themselves use the term, the Japanese doctors involved in these ongoing examinations—in California, and in Japan as well—have become increasingly sensitive to the reality of "leukemia of the soul." As this book goes to press in mid-1997, the most recent findings of researchers reinforce this relatively new focus on the enduring trauma of having survived nuclear violation. An emerging area of concern, for example, involves a higher incidence of hypertension among survivors, possibly related to stress. At the same time, the books have not been closed on either the long-term or hereditary effects of bomb-related radiation. Since radiation sensitivity is greatest early in life, the survivors who are presumably most vulnerable to delayed damage—those who were under twenty in 1945—are only now entering their autumn or winter years. And although statistically non-significant, small increases in genetic damage have been noted and further studies have been proposed. The story is far from closure.

As Professor Sodei's chronicle reveals, moreover, there are stories within stories here.

Some survivors, for instance, appear to have put their hellish experience behind them with remarkable equanimity. For others, the pain remains so deep, even shameful, that they have refused to speak about their ordeal, or to take advantage of the medical examinations which the political struggle of the 1970s succeeded in obtaining. What makes *Were We the Enemy?* such a broadly demystifying account is that no single voice speaks for "hibakusha." By the same token, there is no single "atomic-bomb experience"—nor it is possible, after encountering these diverse individuals, to speak simplisticly of a "Japanese American" personality or experience. These groups are neither homogenized nor romanticized. Their experiences and responses vary greatly. Some survivors of the bombs harbor anger and hatred toward the bombers, others express forgiveness, still others have to all appearances suppressed memory of the trauma of a half century ago almost to the vanishing point. Similarly, while most Japanese Americans, including those trapped in Japan by the outbreak of the war, may have felt primary loyalty to the United States, more than a few identified with the Japanese cause—and small wonder, given the prejudice they experienced in the States. We encounter people torn between two countries and cultures, helpless in the cauldron of war. Stereotypes fall away, and we are left with an intensely human experience.

Professor Sodei's great accomplishment lies in enabling us to see these American hibakusha as individuals, and through them to see both the atomic bombs and the Japanese American experience in new light. He has brought exceptional credentials to his task. Long active in the Japanese peace movement, he is deeply versed in the literature about Hiroshima and Nagasaki. Trained as a graduate student at UCLA, he is one of Japan's leading analysts of American politics—and, at the same time, has devoted many years to observing the Japanese community in California as both outsider and insider. Because many of the survivors we encounter in *Were We the Enemy?* are more comfortable speaking Japanese rather than English, Professor Sodei has been able to engage them in frank and natural exchanges that would never have taken place relying on English alone.

And he has the human touch. When all is said and done, that is what this book is about.

John W. Dower

Notes

1. I am indebted, in the comments that follow, to Dr. James Yamazaki for both technical information on research pertaining to the effects of the bombs and his personal observations concerning the bomb survivors in California. Dr. Yamazaki's account (with L. B. Fleming) of the effects of the bombs on children— *Children of the Atomic Bomb: An American Physician's Memoir of Nagasaki, Hiroshima, and the Marshall Islands* (Durham, NC: Duke University Press, 1995)—can be read with great profit alongside the present work.

Introduction

Much of this book was written as a series of reports from the field for *Ushio*, a Japanese monthly journal, between July 1977 and September 1978, under the title "The Saga of Hibakusha (Atomic-Bomb Victims) in America." When these reports were compiled in a book later in 1978, it caused me to reflect on why it had taken me so many years to awaken to the story of the American survivors of the atomic bomb.

I'd had numerous opportunities to become aware of the American hibakusha. I had spent the early 1960s as a graduate student in Los Angeles, where there were so many Japanese Americans with Hiroshima roots that the area was jokingly referred to as Rafu (L.A.) County of Hiroshima Prefecture. Back then, there were many Issei (first-generation Japanese) still very much alive, who would often tell stories of their experiences in the relocation camps during World War II. There were hibakusha among my Nisei (second-generation) acquaintances, and I was even aware that there was an atomic bomb memorial service every summer at the local Buddhist temple. But I didn't pay any serious attention at the time.

In the spring of 1970, I made a trip to New York in preparation for the American exhibition of the "Hiroshima Murals," a series of powerful paintings of the aftermath of the atomic bombing of Hiroshima by Iri Maruki and Toshi Maruki. While there, the writer and editor Norman Cousins remarked to me, "The atomic bomb is a sad link between Japan and America." Looking back, I see that the hibakusha who were born in America and experienced the bomb in Hiroshima were the very embodiment of that link, but it would still be some time before they entered my line of sight.

Later that year the "Hiroshima Murals" exhibit opened in New York, and I accompanied the Marukis on their first trip to the United States, traveling to various cities to lecture and show a film about their paintings. In the Little Tokyo section of Los Angeles, we spoke at the Pioneer Center, a hall dedicated to the Issei generation. After the presentation, a woman approached me to say, "I am actually a Hiroshima hibakusha." I believe she is one of the people whose stories I later heard and presented in this book, but that day we were in a rush to get to the next event at UCLA, and we had no time to speak with her at length. Once again, I missed a chance to connect with the American hibakusha.

It wasn't until the spring of 1975 that I finally managed to awaken to their situation. On a trip to Los Angeles, I met Kaz Suyeishi, who told me about the Committee of Atomic Bomb Survivors, which was active in California. Later that year, I traveled around the world to report on the antinuclear movement, a journey that began with a meeting of hibakusha in San Francisco. The voice of the American hibakusha finally began to reach my ears.

The present book began to take shape in 1976 when, at the suggestion of the editor of *Ushio*, I returned to California as the first step in an ambitious plan to conduct a survey of all the hibakusha in the United States. It proved a most difficult task, for the situation in which the individual hibakusha existed was so severe, and their problems were such a source of pain, that most remained reluctant to even meet with me. Even when they agreed to see me, it was hard to get them to speak from the heart. I reported on this trip for *Ushio* in September 1976, under the title, "Appeal from Hibakusha in America." The report created quite a stir in Japan, but I felt a deep dissatisfaction with the limitations of my work. The very difficulty of reporting this story was an indication of the size and seriousness of the problems confronting the hibakusha. I decided I would take on the challenge and try to find a way into the heart of the story.

I brought to the task a long-standing interest in two subjects that form the core of this inquiry: the history of Japanese immigration to the United States, and issues surrounding the atomic bomb and nuclear energy. During my six years of study at UCLA, I had lived in the midst of the large Japanese-American community in Los Angeles and had begun to explore the story of Japanese settlement in California. After my return to Japan in 1965, I began a lifelong involvement in nuclear issues, helping to organize support for the Maruki Museum for the Hiroshima Murals and participating in a Citizens' Committee to Convey Testimonies of Hiroshima and Nagasaki. If I were to follow these two strands of inquiry and engage in further research, I thought I might be able to piece together the jigsaw puzzle of the American hibakusha. I hoped this would also provide a window through which to understand the larger subject of the history of Japan-U.S. relations.

I made five more trips to the United States to conduct research. As it happened, my research coincided with the peak of the public activity of the American hibakusha. I was able to witness firsthand many of the key events, meetings, and hearings in the hibakusha's efforts to gain recognition from the U.S. government, the media, and the public at large.

A revised and updated edition of this book was published in Japanese in 1995, on the occasion of the fiftieth anniversary of the bombing of Hiroshima and Nagasaki. That version was further edited and updated to produce the present volume. This remains the only comprehensive account of the American hibakusha in any language.

The chronicle of the lives and struggles of these people tells an essential and largely neglected story of the nuclear age. The number of American hibakusha—some three thousand were in Hiroshima when the bomb was dropped, and between eight hundred and a thousand survived and later returned to the States—may be small in relation to the total number of survivors of the atomic bombs, and their situation is certainly a special case. But I have long believed that the larger truths of history reside in the neglected shadows of society.

Despite the decades of concerted effort that are chronicled in this book, the very existence of the American hibakusha has yet to be acknowledged by the U.S. government. The media and the public at large remain largely oblivious to the fact that thousands of Americans were killed and hundreds of other Americans were left with physical and psychic scars from the bombs dropped on Hiroshima and Nagasaki.

The refusal to countenance this historical fact and admit it into the "story" of the atomic bombs can be traced, in part, to the fact that the American victims of the bombs were *Japanese* Americans. They are therefore virtually indistinguishable, in the public imagination, from the World War II "enemy," despite the fact that most of the American citizens in Hiroshima at the time of the bomb were adolescents or young adults. The title of this book raises the question: Why are the American victims of the bombs still counted among the "enemy" half a century after Hiroshima and Nagasaki?

But there is another and perhaps more significant reason that the American victims of the bombs are nowhere to be found in the pages of history. This is that the atomic bomb, as terrible a weapon as it turned out to be when used on the human populations of Hiroshima and Nagasaki, was portrayed and widely perceived by the American public as the final blow that ended an even more horrifying war. Any historical fact that diminishes the presentation of the atomic bomb as liberator from the war represents an uncomfortable reality that is difficult to reconcile. This has been especially true of the fact that a significant number of the victims of the bomb were Americans, many of whom continue to suffer their private anguish.

In the years since this book was first published, the Cold War has come to an end and the worldwide situation regarding nuclear weapons has changed dramatically. Still, as the recent controversy over the abortive attempt to include displays about the victims of the atomic bomb in the *Enola Gay* exhibit at the Smithsonian's National Air and Space Museum revealed, it remains difficult to reconcile the uncomfortable facts of history and the mythic status of the atomic bombs. The issues that were at the heart of this book twenty years ago remain very much alive today.

This book would not have been possible without the cooperation of the American hibakusha, who shared with me the stories of their experience of the atomic bomb and the lives they have lived in the decades that followed. In particular, I would like to acknowledge my debt to Kanji Kuramoto and Kaz Suyeishi, two dedicated leaders of the hibakusha effort to gain recognition in the United States, who assisted me in innumerable ways over the span of many years.

I would also like to thank a number of people who helped make this English edition of the book possible. The late Barbara Reynolds read the entire manuscript and provided many editorial suggestions. Tadatoshi Akiba, a former professor of mathematics at Tufts University and now a member of the Lower House of the Japanese Diet, encouraged me to produce an English edition and helped organize a pool of translators who contributed their time and talents to this project as a labor of love. John W. Dower, professor of history at the Massachusetts Institute of Technology, provided tireless support that went far beyond the bounds of ordinary friendship. Thomas Noguchi, former Coroner–Medical Examiner of Los Angeles County, educated me on many technical aspects of exposure to radiation. James Yamazaki, a physician in Los Angeles who has been involved in the periodic examinations of the hibakusha, reviewed the manuscript from the perspective of current knowledge of the physiological and psychological effects of the bomb. Mark Selden, professor of sociology at Binghamton University, found merit in the book and served as a virtual manager between the author and the publisher. Finally, John Junkerman's editorial skill shaped a book out of the original translation.

one

From Hiroshima, Back to Hiroshima

On the morning of August 6, 1945, the city of Hiroshima appeared bright in the summer sunshine when it was sighted from the cockpit of the B-29 bomber, *Enola Gay*. Moments later the city would be engulfed in a hellish conflagration.

Approximately three thousand American citizens of Japanese descent, known as Nisei (or "second generation"), were living in Hiroshima on that day, but there is no evidence that the crew members of the *Enola Gay* had any knowledge of their presence. There is no record of this ever being discussed among military personnel on the island of Tinian, the launching base of the *Enola Gay*, or during planning meetings in Washington, where the decision to drop the atomic bomb on Hiroshima was made. The Nisei in Hiroshima were truly "forgotten Americans."

The following are the stories of seven of these American hibakusha.

At the time of the bombing, Sadako Obata was twenty years old and five months pregnant.[1] She lived in the Nishikuken-cho district of Hiroshima, just five hundred meters from the hypocenter of the atomic bomb. Her husband was in the military and stationed at a barracks in the city, so Sadako was living with her older sister and a year-old nephew.

Sadako was in the kitchen after breakfast when a blinding flash struck her eyes. She heard a tremendous noise and lost consciousness. When she came to, she found herself pinned to the floor. A piece of wood had skewered her cheeks.

"Help!" Sadako heard her sister cry out, followed by screams from her nephew. Despite unbelievable pain, she succeeded in pulling the splintered wood from her cheeks. Devouring flames were approaching and all she could do was flee—leaving her sister and the child behind.

Sadako spent the night on a riverbank near Yokogawa and, overcome by thirst, drank the black rain that had started to fall. The following day, she stayed in Yokogawa. It was not until the third day that she was carried by truck to her father's house in the village of Kabe, where at last the wound on her face was treated (a large scar remains). Her husband, who had been exposed to the bomb at the military parade grounds but suffered no apparent injury, came to Kabe and found Sadako there. About ten days later, he began spitting up blood and soon died.

Sadako's sister-in-law, a Red Cross nurse who had dressed her wound, also died. Sadako was left to carry her husband's and sister-in-law's corpses into the hills and cremate them. Her older sister had somehow crawled out of the ruins of her home and managed to make her way to an aunt's house, only to die there shortly afterward. Sadako learned of her fate by chance, when she visited the aunt later on the very day her sister died. There she learned that her father-in-law and another sister-in-law had also perished. She was literally surrounded by death.

Had Sadako remained in the United States, she would not have experienced this living hell. Born in the small southern California town of San Juan in 1925, one of eleven children of a farm family, she came to Japan with her parents at the age of fourteen. The purpose of the trip was to visit Sadako's grandmother, who was bedridden, but Sadako's parents apparently wanted the little girl, who was too small to help with work on the farm, to be brought up in Japan. They returned to the States shortly afterward, leaving Sadako behind. When the war broke out, she was attending a girls' school with her older sister, who had come to Japan earlier. She later learned that her parents and all her brothers and sisters had been sent to the Poston Relocation Center in Arizona.

To Jane (Yoshika) Ishigame, a seventeen-year-old student at the time of the bomb, Japan had always remained an alien country.[2] If her grandfather had not decided to go back to Hiroshima after his retirement in 1938 and taken her with him, she would have spent her youth in Jerome Relocation Center in Arkansas after the outbreak of the Pacific War. That's where her mother (born in the town of Saka in Hiroshima Prefecture) and her five brothers and sisters ended up. Her father was a Nisei from Hilo, Hawaii, who had spent his adolescence in Japan and then returned to the United States. At the outbreak of the war, he was engaged in farming in

Fresno, California, where Jane was born. He was separated from his family and confined in a detention camp in Santa Fe, New Mexico.

Jane's older sister, Irene, also accompanied their grandfather to Japan. The actual purpose of their trip was to obtain medical treatment for Irene, who was ill. They only intended to stay temporarily, until Irene recovered, but war broke out in the meantime. Their parents could no longer send money, and life was difficult for Jane, Irene, and their grandfather. School life for a girl with U.S. citizenship was hard to bear. All of a sudden, America became an enemy country, and the teachers and other students hated Jane because she was an American. She wanted the war to end as soon as possible, but could she have imagined it would end in such a tragic way?

In her fourth year at Yasuda Girls' School, Jane was mobilized to work at a needle manufacturing factory run by the Kowa Sewing Machine company, at the foot of the Yokogawa Bridge, about a half mile from ground zero. There had been an air raid warning early in the morning. Jane had just returned to work at her machine after the "all clear" had sounded when her surroundings lit up suddenly, as if a flash had been set off. That's all she remembered. When she recovered consciousness, she saw dead friends all around her. Only four or five managed to crawl out from beneath the crushed roof. Jane fled, as large drops of black rain pelted her back. Somewhere, she sat beside a hut with a friend until evening. She remembered being fed the juice of cucumbers from a nearby field.

The two girls were taken by cart to what seemed like a rural elementary school, but she did not remember where it was or how many days she spent there. Many people around her were dying. Then the father of a friend who lived in a neighboring village happened to come by, searching for his own daughter. He saw her name tag and the badge on her blouse. "Why, this is the Ishigames' daughter!" he declared, and took her back with him. When she returned to her home, everyone was discussing her funeral.

In August 1945, Kaz Suyeishi was in her nineteenth year, the year known to Japanese as *yakudoshi* and traditionally regarded as unlucky.[3] She had been brought up so thoroughly Japanese that it was natural for her to refer to this somewhat archaic belief. For although she was born in Pasadena, California, where her father was working, the whole family returned to Japan within a year of Kaz's birth. During the war, Kaz completed advanced studies at a girls' high school and, in 1945, was working as a member of a girls' volunteer corps at the Kannonmachi branch of Mitsubishi Heavy Industries.

On the day of the atomic bombing, however, she happened to be resting at home, where she was recovering from a fever. At the moment of the flash, Kaz covered her eyes with her hands and threw herself flat on the ground. Something fell from above and struck her very hard across the hips. Everything became terribly quiet. All was gray and gloomy. Then she saw her mother appear with her hair in disarray, followed by her father, who was bleeding from his head and arms. In the afternoon, her younger brother made his way back from Hiroshima First Middle School. His white shirt had been turned chocolate brown by blood. When the black rain started to fall, she and her younger brother rushed to the air-raid shelter, crying in panic, "Watch out! The enemy is spreading gasoline!" They were literally frantic.

Afterward, Kaz's gums began to bleed and she developed a high fever. No one at that time knew that her symptoms were caused by radiation. She strained herself taking care of her father and brother, who were both close to death. After they recovered, Kaz was bedridden herself for half a year. It was only in later years that her bond with the United States was revived.

There were many among the American-born hibakusha who, like Kaz Suyeishi, had little reason to be conscious in their daily lives that they were American.

At the time of the bomb, thirteen-year-old Akira Furuta (not his real name) was in his second year at Hiroshima Second Middle School.[4] He described himself as a militaristic boy, eager and active in practicing the martial arts of judo and kendo, and quite prepared to stand up to American soldiers once they invaded Japan. Born in Portland, Oregon, Akira remembered coming to Japan with his mother in 1939, when he was in the first grade of elementary school. As far as his outlook was concerned, Akira was Japanese to his fingertips. The only subject he failed in Middle School was English. It was his ambition to enter the Naval Academy.

On August 6, Akira joined the ranks of other classmates at the East Parade Ground behind Hiroshima Station. Being a well-disciplined soldier-to-be, despite the sultriness of the summer morning he was properly attired in his uniform jacket. The bomb blast slammed him down on the dirt of the parade grounds, which had been plowed to plant potatoes. When he opened his eyes, he seemed to be right in the middle of a fire, but thanks to his jacket, only the left half of his face and the back of his hands were burnt and swollen.

Akira ran back to his house in the Furuichi-machi district, where his mother dressed his burns with medicine she had brought back from the

United States. The next day he made his way to his school, where nothing remained but burned ruins and the skeleton of the vice principal who had been on duty that particular day. About eighty percent of his classmates died by the atomic bomb.

Two weeks later, a woman who lived two doors down from Akira died. His mother told him to go to the woman's funeral, but he refused because he did not want to see a dead person. It was only later that Akira learned that this woman was actually his real mother. When he was only a year old, his father had died and Akira had been adopted by a couple who lived in Tacoma, Washington, and who were originally from Hiroshima. His stepfather had returned to Japan on the last ship before the outbreak of the war, and he had registered Akira as his own child. Akira had not been told that his birth mother had also come back to Hiroshima before the war with her other children or that she was living in the neighborhood. The children Akira thought of as his playmates had actually been his older brother and sister.

Jack (Motoo) Dairiki was born in Sacramento in 1931.[5] In August 1941, he came to Japan with his father to visit his sick grandmother. Shortly after they arrived, the war broke out. Jack attended Okukaita Village Elementary School and then a municipal technical school, where the students were mobilized to work every day. Jack was working at a Toyo Industries plant in Mukainada, operating a German-made lathe in the production of rifles. On August 6, however, his entire class was scheduled to go into the city to help with the demolition of buildings to create firebreaks.

The departure of Jack and his companions was delayed by a breakdown on the Sanyo Railway line. While they were lined up in the factory yard waiting to leave, Jack saw an American plane over the city of Hiroshima three miles away. It was the first B-29 he had ever seen. Thinking that even if a bomb were dropped on the city it wouldn't reach where he was, he nonetheless started moving toward a shelter. In that instant there was a flash. The light given off was far brighter than the sun.

Jack instantly threw himself flat on the ground, a conditioned reflex he had acquired through air-raid training. After a while, he tried to look around, but he could not see clearly through the blanket of dust. He got to his feet and looked toward Hiroshima. He witnessed a gigantic mushroom cloud soaring into the sky. Over three years later, when he returned to the United States, Jack still carried a vivid picture of that mushroom cloud deeply imprinted on his memory.

Kaname Shimoda was also a Nisei from Sacramento.[6] As a student at the First Municipal Technical School, he had been mobilized to work at the Hiroshima Medical School. It was there that he experienced the atomic bomb. Making his way past the foothills of Hijiyama, he struggled back to his home in Senogawa-cho. On his way, he saw the mushroom cloud from a pass through the hills. He recalled that the lower part was blazing red in whirls of flames. He guessed he must have been somewhat calm at the time, because he can remember that he thought it was pretty. The right half of his body was badly burned, and several days after he reached home Kaname lost consciousness. He remained in a coma for three days.

Judy (Aya) Enseki was exposed to the bomb while she was on her way from her house to a nearby field.[7] Thanks most likely to the parasol she happened to be carrying, she hardly suffered any burns. Her two-year-old son, who had been born in America, was not hurt either, although he was splattered with mud. Perhaps that was why Judy remained calm enough to observe the mushroom cloud. "It was huge," she recalled, "with a pure white rim. But in the center, the glaring red smoke was seething like a storm."

Judy was born to a farm family near Delano, in central California, the fifth of eight children. In early 1942, before she was moved to a temporary assembly center in Fresno, Judy met and married a young Kibei Nisei (a Nisei who had returned to the States after spending time in Japan) who had been brought up in Hiroshima. It was sympathy rather than love, she says, that made the nineteen-year-old "country girl" decide to get married.

In September of 1943, two months after she had given birth to a child at the Manzanar Relocation Center, Judy received notification that she and her husband were to be allowed to return to Japan aboard an exchange ship. The arrangement was the result of good connections her brother-in-law had made while working at the Japanese embassy. Judy had no desire to go to Japan; for her, America was home. But she could not defy the traditional Japanese teaching that a wife should obey her husband.

So Judy left Manzanar with her child to accompany her husband, who renounced his American citizenship. They made stops at the Poston and Gila camps, both in Arizona, and headed on to New York. From there, the *Gripsholm*, a Swedish vessel, took them to Goa, India, where they were transferred to the *Teia maru*, a Japanese ship. Eventually, by way of Java, Sumatra, Singapore, and Manila, Judy, her husband, and their child reached Hiroshima on a military ship in the spring of 1944. Shortly afterward, her husband was drafted and sent to Manchuria, where he was taken prisoner by Soviet troops. Judy's long voyage turned out to be nothing but a journey to disaster in Hiroshima.

Why did so many Japanese Americans happen to be living in wartime Hiroshima? The answer can be traced to the fact that, during the prewar period, more Japanese emigrated to the United States from Hiroshima Prefecture (where the city of Hiroshima is located) than from any other of Japan's forty-seven prefectures. According to Japanese government statistics for the end of 1936, a total of 73,763 overseas Japanese had come from Hiroshima. Of these, 26,403 were living in Hawaii, 22,604 in the continental United States, and 11,956 in Brazil.

There is no simple explanation as to why emigration from Hiroshima had been so popular, compared to other parts of Japan. Hiroshima suffered from a scarcity of arable land and serious overpopulation, but it was nonetheless in far better condition than Okinawa, for example, with its reputation as the prefecture of "always a million too many people." Hiroshima was also far better off than the poverty-stricken prefectures of northeastern Japan. But there was a sequence of circumstances that worked together to provoke such large numbers to emigrate from Hiroshima.

The first of these circumstances was the decision in 1884 of Sadaaki Senda, governor of Hiroshima Prefecture, to build the port of Ujina in an effort to promote regional development. To build the port, Senda ordered the construction of a landfill in the Inland Sea off of Nihoshima Village. Fishermen, deprived of the fishing grounds where they had harvested seaweed and oysters, opposed the plan with mass demonstrations where they waved bamboo spears in the tradition of peasant revolts, but their efforts were in vain.

The following year, an agreement on immigration was reached by the Meiji government and the king of Hawaii. It must have been a source of great relief to those who lost their livelihood because of the landfill: Of the 945 emigrants who boarded the first ship bound for Hawaii, the *City of Tokyo*, 220 came from Hiroshima Prefecture. Of these, 150 were residents of Nihoshima. In the relatively short history of Japanese emigration, as many as 5,300 people emigrated from Nihoshima.[8]

Although the Japanese arrivals were called "immigrants" when they reached Hawaii, they were actually migrant workers on three-year contracts to work in the sugarcane fields. As their contracts expired, many of these immigrants moved on to the United States mainland. A song of that time reveals the indecision many of them experienced: "On to America or back to Japan? This Kingdom of Hawaii is the crossroads."

By the time Hawaii was annexed to the United States in 1898, the number of Japanese emigrants had increased dramatically—coming to Hawaii, directly to the mainland, or on to the mainland after a stay of a few years in Hawaii. The annual figure for emigrants from Hiroshima Prefecture ranged as high as seven thousand. By the end of 1907, more than thirty-six thousand had emigrated.[9]

After the exodus of the Nihoshima fishermen, there followed an emigration of farmers, whose main crop had been cotton. A great influx of imports forced these farmers to cut back production of cotton, a crop that had ranked third in Japan, after rice and barley. This dealt a decisive blow to farmers whose economic position had already suffered as a result of the taxation policies of the new Meiji government. The majority of Hiroshima emigrants originated from the city of Hiroshima itself, the downstream areas of the Ōta River, and the western shore of Hiroshima Bay—all of which were heavily engaged in the cultivation of cotton.

Poverty and need, however, were not in themselves sufficient motivation to leave Japan. An additional stimulus was provided by the success stories of those who had ventured abroad in the early years of emigration and who returned in triumph after years of hardship abroad. Those who stayed behind became caught up in a kind of emigration fever. There was a saying in those days, "Become a soldier or go to America!" For poor young men in the countryside, military service and emigration to the United States were among their limited choices, each carrying its own set of challenges.

By the end of 1915, some 59,591 people had emigrated, most of whom were residing in the United States or Hawaii.[10] Like those who had earlier emigrated to Hawaii, many who went to the mainland also engaged in agricultural labor. Most were farmers to begin with, and since they could not speak English and had little knowledge of American culture, the easiest route for them was farming. Oftentimes they purchased barren land cheaply and, with tremendous effort, started their own farms.

After agriculture, work in railroad construction and in the mines were the most common occupations for immigrants. Japanese workers were considered diligent and ingenious, and they were especially welcomed after the Chinese Exclusion Act of 1882 put an end to the immigration of low-wage Chinese workers. As the immigrant population increased, Japanese in cities like San Francisco and Seattle ventured into such service industries as hotel management, restaurants, grocery stores, and laundries. Gradually, Japanese spread from California and Washington to many other parts of the country.

Japanese immigrants were subject to racial discrimination from the very beginning, especially on the West Coast where Asian immigration was concentrated. Ever-growing numbers of Chinese and, later, Japanese immigrants provoked public protest, especially from organized labor. The newcomers competed in the labor market by working for lower wages, and they were from unfamiliar cultures and thus easy targets. Politicians took advantage of an intensifying mob psychology and further aggravated the situation.

After they succeeded in suspending Chinese immigration in the 1880s, these politicians and others set as their next goal the restriction of Japanese immigration. After Japan's victory in the Russo-Japanese War of 1904–1905, Japan was portrayed as a potential military threat. The cry of "Yellow Peril!" poured oil on the flames of the anti-Japanese movement in California and elsewhere. Violent incidents had taken place earlier, but in October 1906, San Francisco municipal authorities for the first time imposed official sanctions that prohibited Japanese children from attending American schools.

The sanctions were lifted half a year later through the mediation of President Theodore Roosevelt, but the Japanese government was forced to accept an executive order that prohibited further Japanese immigration from Hawaii to the mainland. This executive order was just the beginning of a series of actions by which the United States sought to discourage Japanese immigration and diminish its effects.

The next steps in this direction were the so-called "Gentlemen's Agreement" memoranda exchanged between the two governments in Washington and Tokyo in 1907 and 1908. Under this agreement, immigration was permitted, with few exceptions, only to parents, spouses, and children under the age of twenty, of Japanese who were already residing in the United States.

The period of Japanese immigration that followed is known as "yobiyose," or immigration by family sponsorship. The new restrictions led to a decline in the number of Japanese living in the United States. Many of the Japanese had left their homeland with the intention of working several years in the United States and then returning to Japan with substantial savings. Once anti-Japanese sentiment became widespread, many simply chose to leave. According to a history of U.S.-Japanese cultural relations, the Japanese population in the United States increased by about ten thousand annually after 1904 until it reached 103,000 in 1908. But one year later it had dropped to 98,700.[11]

Many of the Japanese who immigrated prior to the "Gentlemen's Agreement" were single men, or men who had left their wives and children behind in Japan. For married men who chose not to go back to Japan, the goal was to summon their wives and children and establish a home in America. The dream for single men, on the other hand, was to return to Japan temporarily, to seek out a wife, and then to return to the United States. (Both ends of this journey represented a "return," which indicates the complex state of mind that beset those who lived between two countries.)

For those who could not afford to return to Japan to find a bride, the solution was to pursue a "picture marriage." According to this practice, a marriage was agreed upon by the couple after an exchange of pictures. The bride was entered in the family register of her bridegroom in Japan (the method for officially establishing marriage in Japan), and then set out for the United States alone. Americans tended to ridicule such marriages and call the bride a "mail-order wife," but the fact was that in Japan the practice of arranging marriages through an exchange of pictures was nothing new. Japanese did not consider it odd if a bride and bridegroom did not even meet until the day of the wedding. Picture marriage was simply a custom imported across the Pacific.

It is not known exactly how many "picture brides" entered the United States through this convenient and relatively inexpensive arrangement, but the Immigration Office of the Port of San Francisco recorded a total number of 5,907 arriving during the nine years from 1912 (879 brides) to 1920 (697 brides). During the same years, 1,018 picture brides arrived through the Port of Seattle.[12] The total never exceeded ten thousand because, sensitive to American public opinion that marriage between people who had never met was somehow primitive, the Japanese government banned the emigration of picture brides in March 1920. Picture marriages produced many human dramas and tragicomedies that make an interesting chapter in the history of Japanese immigration. It is highly probable that a good number of those Nisei who were living in Japan during World War II were the children of picture marriages.

The 1924 revision of the U.S. Immigration Act, known as the "Asian Exclusion Act," brought an end to immigration from Japan. This revision was aimed at limiting the rush of immigrants to the United States from throughout the world after World War I. However, the law was especially intended to curtail immigration from Asia. Since at that time the Japanese were the only Asians permitted to immigrate to America, albeit with restrictions, the revised law became known in Japanese history as the "Anti-Japanese Immigration Law."

In November 1922, two years before immigration was banned, the U.S. Supreme Court ruled in *Ozawa v. United States* that Japanese had no right to be naturalized as U.S. citizens. First-generation immigrants (or, "Issei") thus became "ineligible aliens." And the 1924 law prohibited the immigration of all "aliens ineligible to citizenship," virtually outlawing the introduction of new blood into the Japanese immigrant community.

Japanese immigrants suffered discrimination in economic as well as social matters. In 1913, the California Legislature had passed the bill prohibiting aliens from owning land. Then, in November 1920, a referendum was placed on the ballot that denied Japanese even the right to *lease* land. In the course of their campaign, supporters of the bill appealed to racial sentiment with the slogan, "Keep California White." The Japanese countered with their own slogan, "Keep California Green."¹³ Japanese immigrants believed that it was their labor that had enabled California agriculture to develop as quickly as it had. Nevertheless, the "anti-Japanese land bill" passed by an overwhelming majority.

Similar bills were soon passed in Washington, Nebraska, Texas, and Nevada, driving increasing numbers of Japanese off the land. Some came up with the idea that they could purchase and cultivate land in the name of their Nisei children, who were American citizens by birth. This route, plausible as it may have seemed, was blocked when the Supreme Court ruled that the purchase of any land in the name of a child under twenty-one years of age would be considered a violation of the land law. The only livelihood left to Japanese farmers was to work on a contract basis on American-owned farms. Since more than half of the Japanese in the United States had been engaged in agriculture, the "anti-Japanese land law" dealt them an incalculable blow.

Although denied the right to be naturalized or to own land, the Issei continued to make efforts to establish a foothold and to create a distinct subculture where Japanese influences predominated. The children of the Issei were American citizens, who were increasingly assimilated into American culture. But the first generation still carried the culture and spirit of Meiji Japan. They lived in the United States, but their hearts still "turned toward Japan."

The desire to communicate in their native language with their English-speaking children led the Issei to establish Japanese-language schools in many immigrant communities. Nisei children spent their after-school hours and Saturdays learning Japanese. Prefectural associations were formed in many areas, and these associations or the community as a whole organized traditional Japanese events, like New Year's celebrations, the observance of Obon (the Buddhist All Souls' Day), sports days, and summer picnics.

The Issei taught "Japan" to their children, but many also wanted to have them raised in Japan. Large numbers were able to fulfill this desire, although there may have been more than one reason for doing so. One Nisei believes that his parents sent him to his grandparents in Japan because they were so hard-pressed in their daily working lives that they had little time to devote to looking after their children.

More Nisei girls than boys were sent to Japan for schooling. Many Issei parents were dismayed at the prospect of their daughters becoming Americanized, and hoped that they would acquire the "beautiful virtues" of Japan. Indeed, when arranging marriages for their sons, many Issei specifically listed "education in Japan" as a requirement for prospective daughters-in-law. For this reason, not a few Nisei girls were sent to study in girls' schools in Hiroshima, institutions that took pride in educating girls to be "good wives and wise mothers." A former teacher at Hiroshima Jogakuin reported that it was widely believed that a girl could "marry into a family one class higher" back in the United States if she could add "excellent education in Japan" to her personal profile.[14]

The Great Depression also increased the appeal of sending children back to Japan to be educated. In the 1910s, marriage rates among Japanese in the U.S. had reached a peak, boosted substantially by the practice of picture marriage. As a result, the birth rate rose dramatically. At the outbreak of the Great Depression, many of these "baby boom" Nisei children were entering junior and senior high school. The Issei parents saw little hope for their children's future in the America at a time when college graduates had trouble finding employment. Not a few Nisei were sent to Japan by parents who sincerely believed that their children would have a better chance of landing jobs in Japan after receiving an education there. The Issei also reasoned that their children should be brought up in Japan from an early age because, unable to imagine a future for themselves in the U.S., they would eventually go to Japan anyway.

The Japanese Ministry of Foreign Affairs estimated that the number of Nisei from both the U.S. mainland and Hawaii who were living in Japan for family reasons or for education reached almost 30,000 as of January 1929.[15] Of these, 4,805, or sixteen percent, were living in Hiroshima Prefecture. Their ages ranged from one to thirty, but 3,803, or some eighty percent, were attending elementary and middle schools. According to the same statistics, 11,312 Nisei in the United States, excluding Hawaii, had parents who came from Hiroshima, while Nisei residing in Hiroshima numbered 3,404. When 2,759 of the latter group were asked in 1929 whether they wanted to go back to the United States, only 755 answered yes, while 2,004 said no. In other words, seventy percent expressed no desire to return.

After the 1924 revision of the Immigration Act prohibited Japanese from immigrating, only Nisei possessed the right to enter the country without restriction. A book published in 1929 about Hiroshima immigrants in the United States emphasized, "From the viewpoint of the development of the Yamato race overseas . . . some measures must be urgently taken to encourage the Nisei in Japan to come back to the U.S."[16]

The foreign ministry survey was made that same year, twelve years before Pearl Harbor. In the intervening years, how many Nisei returned to the United States? A history of the Japanese in America, published in December 1940 by the Association of Japanese Americans in San Francisco, states: "As a result of a nationwide movement that was started around 1935 to encourage Nisei educated in Japan to return to the United States as the only real successors to the Issei, it is estimated that about ten thousand Nisei have returned at the present time." This statement is qualified, however, by the observation that "around twenty thousand Nisei are believed to still be in Japan."[17]

How many of the latter were living in Hiroshima in 1940? No statistics are available, but if we assume that the sixteen percent of the total that prevailed in 1929 remained consistent, we get an estimate of around 3,200 for the number of Nisei in Hiroshima. Most of these would have been living in and near the city of Hiroshima itself.

What did the United States mean to these Nisei? Legally, it was their homeland. According to U.S. law, which embraces the principle that country of birth is the basis of nationality (jus soli), anyone born within the territory of the United States, including Hawaii, is automatically an American citizen, even though his or her parents may not be (children of American parents are also U.S. citizens regardless of where they are born). On the other hand, in Japan family bloodline through the male line determines nationality (jus sanguinis), so every child whose father is Japanese automatically acquires Japanese citizenship, no matter where he or she is born. Because many Issei parents reported the birth of their children to the local government offices in their native villages or towns in Japan, quite a few Nisei possessed dual citizenship. As long as they remained in the United States, this posed no problem. If they came of age in Japan, however, they were faced with two legal duties of citizens: military service and the vote. Under U.S. law, once they engaged in either one of these activities required by Japanese law, they automatically forfeited their U.S. citizenship.

Concerned about this situation, Japanese living in the United States asked the Japanese government to revise the law of nationality. As a result, in 1925, the government established a provision that stated, "When a person with dual citizenship comes of age (twenty years old), he has the right to choose his own nationality."[18] A great many of the Nisei in Japan were too young to make any decision concerning citizenship. But during the years of the war, many of them reached the age where they were forced to confront this serious dilemma.

Possession of dual citizenship implies that a person has two "nations" within himself or herself. What happens when those two nations go to war? For Nisei in Japan, the outbreak of the Pacific War transformed the country where they were living, the country of their parents' birth, into enemy territory. At that juncture, the fact that they had been born in the United States must have seemed an ironic fate that they had to accept. Of course, many were too young to be fully aware of the consequences of that fate but, during wartime, how many Nisei living in Japan could remain untouched by their American birth?

The attack on Pearl Harbor came as a terrible shock to many of the Nikkei, as people of Japanese descent in America are known. While some viewed with foreboding the steady progression of diplomatic and political confrontations between Japan and the United States that preceded the outbreak of hostilities, the majority of the Issei did not want to believe that worsening relations would lead to war. As one person put it, "War between Japan and the U.S. is a dream in a dream. One might fail to hit the ground with a hammer, but no one can fail in this prediction. No Japanese in America should be worried."[19] This judgment was clearly wishful thinking, no doubt influenced by an overwhelming fear that war would make their lives even more difficult than they were already.

The *Tatsuta maru*, which sailed for Japan in August 1941, was crowded with Issei planning to visit their hometowns and Nisei who were looking forward to seeing the country of their ancestry for the first time. Few passengers at that time felt any concern that they might not be able to return to the United States after their visit to Japan. The attack on Pearl Harbor caught them totally by surprise. Not only did it cause great damage to the U.S. Pacific Fleet in Hawaii, it also greatly affected the fate of the Nikkei in Japan and radically altered the lives of some 120,000 individuals of Japanese ancestry who were living along the Pacific coast of the United States.

Notes

1. Interview with Sadako (Obata) Shimazaki, June 21, 1976, Monterey Park, CA.
2. Interview with Jane (Yoshika Ishigame) Iwashika, June 28, 1976, Los Angeles.
3. Interview with Kaz Suyeishi, June 16, 1976, Los Angeles.
4. Interview with Akira Furuta (pseudonym), June 23, 1977, Sacramento, CA.
5. Interview with Jack (Motoo) Dairiki, April 9, 1977, San Francisco.
6. Interview with Kaname Shimoda, September 11, 1977, Los Angeles.
7. Interview with Judy (Aya) Enseki, June 18, 1976, Los Angeles.
8. Soshun Fukagawa, "Ujina," in *Minato kikō* (Travels of Harbors) (Tokyo: Asahi Shimbunsha, 1976), p. 165.
9. Jun'ichi Takeda, *Zaibei Hiroshima kenjinshi* (History of the Hiroshima People in America) (Los Angeles: Zaibei Hiroshima Kenjinshi Hakkōjō, 1929), p. 46. On Japanese immigration generally, see Yuji Ichioka, *The Issei: The World of the First Generation Japanese Immigrants 1885–1924* (New York: Free Press, 1988).
10. Takeda, *Zaibei Hiroshima kenjinshi*, pp. 48–49.
11. Kaikoku Hyakunen Kinenjigyō Iinkai (Cultural Committee to Commemorate the Centennial of Japan-U.S. Relationship), ed., *Nichibei bunka koshōshi* (A Cultural History of U.S.-Japan Relations) (Tokyo: Yōyōsha, 1955), Vol. 5, *Ijūhen* (Immigration), pp. 126–127.
12. *Ijūhen*, p.128.
13. Shiro Fujioka, *Ayumi no ato* (Footsteps of the Issei) (Los Angeles: Ayumi no Ato Kanko Koenkai, 1957), pp. 577–578.
14. Interview with Tazu Shibama, former Hiroshima Jogakuin Girls' High School teacher, May 20, 1978, Hiroshima.

15. These and the following statistics are drawn from Takeda, *Zaibei Hiroshima kenjinshi*, pp. 132–139.

16. Takeda, *Zaibei Hiroshima kenjinshi*, p. 139.

17. Zaibei Nihonjinkai Jiseki Hozonkai, *Zaibei nihonjinshi* (History of the Japanese in America) (San Francisco: Zaibei Nihonjinkai, 1940), pp. 1117–1118.

18. *Zaibei nihonjinshi*, pp. 1108–1109.

19. Fujioka, *Ayumi no ato*, p. 180.

two

Death—and Life— in the Desert

On December 6, 1941, President Franklin Roosevelt issued a full-speed-ahead order for a project with the code name "S-1." This top secret project was the development of the atomic bomb. How ironic it is that Roosevelt's directive was issued on the eve of the bombing of Pearl Harbor.

German physicists had announced the theory of nuclear fission in December 1938, and from that time on many scientists dreamed of using the enormous energy generated by fission in the creation of a "super weapon." To those scientists who fled Adolf Hitler's oppressive regime for the United States, the possibility of such a weapon represented more of a nightmare than a dream. They feared that if the Nazis were to succeed in building an atomic bomb, Hitler would surely conquer the whole world.

This fear was expressed in a letter that Albert Einstein sent to Roosevelt in 1939. Einstein warned that Germany might develop an atomic weapon, and though he did not explicitly recommend that the United States develop an A-bomb, his implication was clear: To prevent Hitler from controlling the world, the United States must precede the Germans in the production of atomic weapons.

But would it be possible? After the war, it was disclosed that German scientists had concluded that it would be impossible to develop the atomic bomb during World War II because of the tremendous resources and manpower required for the task. Initially, the U.S. military was equally skeptical.

America's sense of crisis deepened, however, as the Axis powers gained on the European front. A year after Einstein wrote his letter, the

Office of Scientific Research and Development was established under the direct control of the president. Research on the atomic bomb began in June 1940, centered at Princeton, Columbia, and the University of California at Berkeley. The decision Roosevelt made on the eve of Pearl Harbor took this line of research the next step into development, though he continued to take an experimental approach toward the possibility of building an atomic bomb. "If in six months the project was making definite progress," James MacGregor Burns later wrote, "he would make available all the industrial and technological resources of the nation to bring about a crash production of the atomic bomb."[1]

Roosevelt ordered work to begin on building the atomic bomb in June 1942, under the code name "Manhattan Project." Facing the direct threat of German aggression, British Prime Minister Winston Churchill agreed to pool British atomic-bomb technology. The United States pledged development funds, amounting to $2 billion, and enlisted its own scientists. The threat of a German atomic bomb had driven the United States into atomic bomb production, but undoubtedly the surprise attack on Pearl Harbor accelerated the process and, inevitably, Japan was added to the list of possible targets.

Roosevelt's final go-ahead for the Manhattan Project would affect the fate of all humanity. But another order, one he signed four months earlier in February 1942, altered the destiny of more than 120,000 Americans of Japanese ancestry who lived on the West Coast. This presidential decree, Executive Order 9066, designated the entire western half of the states of Washington, Oregon, and California, and the southern half of Arizona as "military areas," and gave the Secretary of War the authority to remove from this region those who did not have permission, based on "military necessity," to be there. Under this order, all residents of Japanese origin were removed from the Pacific coast and eventually shipped to internment camps in the desert.[2]

Roosevelt's order not only targeted the Issei, whose classification was changed from "aliens ineligible for citizenship" to "enemy aliens"; it also applied to the Nisei, who, as American citizens, should have been granted all the guarantees of the Constitution. In retrospect, this order was clearly a denial of the most basic civil rights, especially for the Nisei. It is equally clear that the decree was racist in character, for the order did not apply to citizens of German or Italian descent, although they too carried the blood of the enemy in their veins. At the time, however, the Supreme Court ruled that the measure was "an exercise of the president's authority necessary for the prosecution of a war" and therefore constitutional.

The forced evacuation of Japanese Americans was a product of wartime hysteria, but the underlying motivation of strong racial prejudice against the Japanese had a long history with deep social roots. For many, the image of the industrious Japanese who had successfully converted barren land and wilderness into fertile farms raised the specter of California being taken over by these immigrants, if nothing was done to prevent it. Pearl Harbor set this bundle of dynamite on fire.

The rationale behind the order was that many of the 120,000 Japanese Americans in the region would almost certainly join the Japanese war effort and carry out subversive activities if Japanese soldiers actually landed on the Pacific coast. This fear became more pronounced as Japanese troops pursued their seemingly unstoppable advance through Asia.

Soon after the beginning of the war, Japanese community leaders and editors and publishers of Japanese newspapers on FBI blacklists had been arrested. They were sent to detention camps in Santa Fe, New Mexico; Missoula, Montana; and elsewhere. Now, public opinion demanded a clean sweep of all Japanese. California Governor Lawrence Olson and State Attorney General Earl Warren (who later became governor of California and eventually Chief Justice of the Supreme Court) made no attempt to control the public hysteria. On the contrary, they joined with congressmen of the western states in pressing the military and the government to remove all Japanese Americans from the region.

Executive Order 9066 was issued on February 19, 1942. The land on which the Issei had built their livelihood was marked "Off Limits," and the Issei were forced to leave their homes and businesses. But where were they to go? Very few had relatives or friends in the Midwest or on the East Coast to whom they could turn for help. The U.S. Army, and later the War Relocation Authority (WRA), proceeded to build camps to house and ultimately confine them. Japanese Americans were interned in temporary holding centers, in the stables of the Santa Anita Race Track, or in barracks on state and county fairgrounds, until facilities could be made available that would accommodate as many as 120,000 people.

It is often suggested that when Japanese Americans presented themselves in orderly fashion and allowed themselves to be "relocated" without resistance, they exemplified the character of the Japanese race. I would argue, however, that their comportment was not a reflection of some ingrained obedience to authority or the manifestation of some national trait of unquestioning loyalty. Rather, many Japanese Americans believed that the only way they could demonstrate their true allegiance to the United States was to accept the executive order without protest and allow themselves to be evacuated. In any case, having experienced the succession of legal and legislative moves that destroyed any illusion of equal rights under the law, the Nikkei had little choice but to comply with a presidential decree enforced by the Army.

In July of 1942, the forced migration of Japanese Americans headed out toward ten facilities that had been hurriedly constructed on federal land in deserts, in swamps, and on former Indian reservations. Michi Weglyn, reflecting on her own experiences as a young Nisei in one of the camps, wrote: "Mindful that evacuees were capable of effecting soil improvements which would turn into postwar public assets, hitherto worthless parcels of real estate were purposely chosen for the WRA campsites."[3]

Although they were named "relocation centers" by the government, these facilities, surrounded by barbed wire and patrolled by armed guards twenty-four hours a day, were little different from concentration camps. And those who lived there called them "camps."

At their peak, these camps held the following numbers of internees:[4]

Poston (Arizona)	17,814
Gila (Arizona)	13,348
Jerome (Arkansas)	8,497
Rohwer (Arkansas)	8,475
Topaz (Utah)	8,140
Granada (Colorado)	7,318
Heart Mountain (Wyoming)	10,767
Minidoka (Idaho)	9,497
Tule Lake (California)	18,789
Manzanar (California)	10,046

Many of these camps were in the desert, where the summers were unbearably hot and in the winter the mercury dropped to numbing temperatures. The internees survived day by day. Their lives were vulnerable to sandstorms, scorpions, poisonous snakes, and other natural hazards. But they more than endured the rigors of their confinement. Just as the government had anticipated, they converted the surrounding wilderness into green lands, if only as means of maintaining their sanity.[5]

While their "fellow countrymen" in the United States were being confined in the camps, what had become of those Japanese Americans who were leading their lives on the other side of the Pacific, in Japan? It appears that many of them were not clearly aware that they were now living in what was considered "enemy territory" back home. Most Kibei Nisei, who returned to the States after the war and were in their forties at the time I interviewed them in the 1970s, were reluctant to talk about those days. The older they were, the more reluctant they were to tell their stories, which may be an indication of the psychological complexity of living in the interstices of two countries at war.

Among those who moved to Japan in early childhood and were still very young when the war broke out, some remember being asked by their teachers to remain after the morning assembly and being interrogated about mail they received from the United States. In many cases, however, they were able to go on with their normal lives. The fact of their American birth was buried deep in their consciousness, and these Nisei behaved like children who had been born in Japan. Influenced by the militarism of wartime Japan, they developed into military-minded young people.

Nisei who were in their teens when the war broke out had to make a more conscious effort to prove their loyalty to Japan, since those around them were more aware that they had been born in America. Japanese could afford to feel some generosity towards the Nisei in the early days of the war, while Japan was winning one victory after another in Asia. But once Japan began to lose the war, many began to treat the Nisei as if they were spies. In response, the Nisei endeavored to demonstrate their faithfulness to the emperor and loyalty to the nation. It was not unusual for the American-born youth of Japan to volunteer for the pilot training corps, the Naval Academy, or the Military Academy.

One woman recalled the bitter experience of being beaten by her teacher while she was in a girls' high school in Hiroshima. The teacher hated her and called her "American." Many suffered financial difficulty because their parents in America stopped sending money to them. A Nisei group was formed, called Himeyuri-kai, or the "Star Lily Society" (in Japanese, the name can also mean the "Japan-America Lily Society"). Separated from their brothers and sisters on the other side of the Pacific, the youths sought comfort in each other's company.

Even more tragic were the cases of parents who were in Japan at the outbreak of hostilities and found themselves unable to return to children they had left behind in the United States. Soon reports reached Japan that Japanese Americans on the Pacific coast had been forced to evacuate their homes and had been locked up in camps. Frightening rumors about conditions in the camps began to circulate. In reality, aside from the humiliation of being confined in the camps, life for the interned Japanese Americans was not unbearable. A minimum standard of living was guaranteed. Even so, early rumors afloat in Japan had it that their jailers were pouring boiling water over the internees at the camps in order to torture them. A woman hibakusha living in San Jose, California, recalled, "I don't know how this kind of rumor got started, but when my father, who was nearly seventy years old at the time, heard such news he became so worried about my brothers that he soon died."[6]

Although torture was not practiced in the camps, the loyalty oath that was imposed on all Issei and Nisei seventeen and older was, in many ways, a form of mental cruelty. The Issei were asked to swear loyalty to a country that refused them citizenship. And the federal government not only demanded that Nisei swear allegiance but that they prove their loyalty by putting their lives on the line in the U.S. military. One democratic right that was still extended to the Nisei was, as Michi Weglyn put it, "the right to be shot at."[7]

As many Nisei saw it, the right to be shot at amounted to an obligation to aim a gun at the homeland of their parents, so the loyalty oath plunged Japanese in the camps into fear and confusion. In Manzanar a riot broke out, and quite a few casualties were caused by the effort to suppress it. Nine thousand Issei and Nisei were labeled "disloyal." Many Issei refused to sign the loyalty oath and demanded to be repatriated. Some Nisei also answered no, and thereby surrendered their citizenship. Between September and October 1943, these resisters were placed in isolation at Tule Lake, a camp located on the border between Oregon and California, where further riots occurred.[8]

In the spring of 1943, the U.S. government constructed another facility in the middle of the desert in New Mexico. Though it was officially called the "Atomic Power Research Institute," this secret facility in Los Alamos, directed by J. Robert Oppenheimer, was the center for development of the atomic bomb. The task of those assigned to Los Alamos was to build a super weapon from uranium-235 that was being isolated at Oak Ridge, Tennessee, and plutonium produced in Hanford, Washington. It was a race against time. Although German troops were beginning to retreat, the nightmare of an atomic bomb in the hands of Hitler continued to haunt the Nobel prize winners and the thousands of young scientists and engineers who were assembled to assist in the project.

In 1944, after their defeat at Stalingrad, the Germans started a general retreat. The possibility that Germany might surrender before work on the atomic weapon was completed grew ever more likely, but the Manhattan Project was nonetheless accelerated.

On July 6, the Japanese forces on Saipan were wiped out. U.S. troops occupying the Mariana Islands began construction of an extended runway and awaited the arrival of a new long-range bomber, the B-29, which was soon to be deployed. A timetable for attacking the mainland of Japan was being drawn up.

On November 1, 1944, B-29s launched from the base on Saipan appeared for the first time over Tokyo. A month earlier, in early October, a small B-29 squadron composed of only fourteen bombers was assembled at Wendover Air Base, located in a desolate corner of the desert in Utah. The commander of the squadron, which was called the "509th Composite Group," was Colonel Paul W. Tibbets. He was the only person who knew that the objective of "Operation Silver Plate," code name for the task assigned to the 509th, was to drop the atomic bomb.

The training exercise assigned to the 509th was a "visual dropping" of a large-sized bomb from a height of six miles, a puzzling assignment to say the least. The plutonium bomb, later nicknamed "Fat Boy," was shaped like a huge bulb, and it took many trials before the task force was familiar with its trajectory. The trial runs often took place over Salton Sea, a lake 280 feet below sea level in the middle of the southern California desert. It is the only lake in the United States that lies below sea level, and the surrounding area is uninhabited. Its northern end opens toward Palm Springs, and its southern end toward the Imperial Valley.

Looking down from an airplane on the Imperial Valley, at first glance the terrain appears blood red. Closer inspection, however, would reveal small swatches of green. Japanese immigrants, working in 120-degree heat, had been cultivating this area since the turn of the century. The green areas in the midst of the desolate valley had yielded the fruit of their labor: Cotton, lettuce, and, especially, melons produced there were the highest quality of any produce grown in the United States. Under the skies that the B-29s were traversing, however, you would not have found a single Japanese. They were all in the camps.

It is short flight from the Wendover base to the Salton Sea, along a route that passes over Las Vegas and the Mojave Desert. In the Owens Valley, situated between the Sierra Nevada mountain range on the east and Death Valley on the west, lay the relocation camp of Manzanar. It is unlikely that the B-29 crew could have seen the camp. Even if they had seen it, they would not have had the faintest clue that the internees in the camp below had hundreds of relatives and next of kin in Hiroshima, who would soon be the victims of the very bomb they were preparing to drop.

Meanwhile, even as the training for the atomic bombing was underway, the defeat of Nazi Germany became inevitable. At this juncture Japan became the sole target. In a memorandum written about two weeks before Roosevelt died, General Leslie Groves, Director of the Manhattan Project, expressed his confident opinion that atomic weapons would be available in time to be used against Japan.

As historian Martin Sherwin wrote:

> He was certain that the weapon would bring the war to a rapid conclusion, thereby justifying the years of effort, the vast expenditures, and the judgment of the officials responsible for the Project. Two bombs would be ready by August 1, 1945. Even before Truman took office, the race for the bomb had already changed from a race against German scientists to a race against the war itself.[9]

On April 23, Groves told Secretary of War Henry Stimson that the target was Japan "and was always intended to be Japan."[10] Once the decision had been made to drop the bomb on Japan, all that remained was to decide which city to drop it on.

Notes

1. James M. Burns, *Roosevelt: The Soldier of Freedom* (New York: Harcourt Brace Jovanovich, 1970), p. 20.

2. Maisie and Richard Conrat, *Executive Order 9066: The Internment of 110,000 Japanese Americans* (San Francisco: California Historical Society, 1972).

3. Michi Weglyn, *Years of Infamy: The Untold Story of America's Concentration Camps* (Seattle: University of Washington Press, 1996), p. 84.

4. *Ibid.*, p. 86.

5. On the internment camps, see also Yamato Ichihashi, *Morning Glory, Evening Shadow: Yamato Ichihashi and His Internment Writings,* edited by Gordon Chang (Stanford: Stanford University Press, 1997); Gary Okihiro, *Whispered Silences: Japanese Americans and World War II* (Seattle: University of Washington Press, 1996); and Page Smith, *Democracy on Trial: The Japanese American Evacuation and Relocation in World War II* (New York: Simon and Schuster, 1995).

6. Interview with Kiyoko Oda, April 22, 1977, San Jose, CA.

7. Weglyn, *Years of Infamy*, p. 136.

8. It should be noted that most of the Nisei agreed to the loyalty oath, and in fact thousands of draft-age Nisei men volunteered to serve in the U.S. military. See Chapter 5 and, more generally, Lyn Crost, *Honor by Fire: Japanese Americans at War in Europe and the Pacific* (Novato, CA: Presidio Press, 1994).

9. Martin J. Sherwin, *A World Destroyed: The Atomic Bomb and the Grand Alliance* (New York: Knopf, 1975), p. 145.

10. *Ibid.*, pp. 209–210 note.

three

Hiroshima: The Target City

Although over sixty major Japanese cities had been targets of U.S. air raids by August 1945, Hiroshima was not among them. Once, in the early morning of April 30, 1945, a single B-29 dropped a small-sized bomb in the middle of the city, resulting in eleven casualties. This was the only bomb attack Hiroshima had experienced.

"Why has Hiroshima not been attacked?" the people of the city wondered. Various speculations ran through the community. The most absurd rumor had it that President Truman's mother was living in Hiroshima. Nobody could be found who had ever seen her, however.[1] According to another rumor, the son of a high official in the U.S. government was imprisoned in Hiroshima as a POW, and this was the reason the United States would not attack the city. There were indeed a number of American POWs in the guardhouse of the military police headquarters in the city. The number of these prisoners was said variously to have been from ten to twenty-three, but the exact number has never been established. Another rumor made the rounds that U.S. bombers had not attacked Hiroshima because many citizens in the city had relatives in the United States, and indeed there were thousands of people living there who not only had relatives but even parents and children residing in Hawaii and on the U.S. mainland.

No one seems to have ever suggested that the reason Hiroshima had been spared was the presence there of a large population of American citizens. These Nisei had been keeping a low profile, trying by every means possible to conceal their American birth. Many American citizens among junior and senior high school students in the city had been mobilized to help the military by doing conscript labor. Nothing was more frightening for the Japanese American than to be treated as a "spy" or a "traitor." To demonstrate his or her loyalty to Japan was the safest route available.

In the spring of 1945, the intelligence department of the Second Imperial Army General Headquarters located in the city of Hiroshima established a "special intelligence squad" and commandeered English-speaking Nisei women to intercept shortwave messages. It is hard to imagine how they could possibly have refused to comply, given the circumstances.

The radio interception group was placed secretly in a back room of a nobleman's mansion, situated in the famous Sentei garden of Hiroshima. Six to seven all-wave receivers were installed, and about twenty Japanese American women were assigned to work three shifts a day. According to Army Lieutenant Colonel Kakuzo Ōya, an intelligence staff officer who was sent to Hiroshima from Tokyo headquarters for the assignment, the purpose was to intercept radio messages from the U.S. Pacific fleet in order to determine its organization and movements. The Nisei operators intercepted messages between the aircraft carriers and the planes. After the defeat in Okinawa in June of 1945, it was vital for Japan to learn when the U.S. Pacific fleet planned to enter the South China Sea.[2]

Irene Ishigame, a squad member who later married and settled in Los Angeles, recalled:

> When I graduated from Yasuda Girls' High School and was working for the Japan National Railways, I received a draft call from the military police. It wasn't voluntary. It was really mandatory. I told them that I didn't understand English because Japan was my second home and that I considered myself to be pure Japanese. They said, "What! If you won't cooperate then you are a traitor." I must have been conspicuous because I was head of the Nisei Federation. All members of this special intelligence team were Nisei. Our job was to listen to the shortwave broadcasts and transcribe the messages. It was a top-secret assignment and we were not even allowed to go outside. It was just like a prison. But I think we served well and showed our loyalty.[3]

Hiroshima had been a military center of Japan since the first Sino-Japanese War of 1894–1895. In the last months of the Pacific War, important military agencies and facilities were concentrated in Hiroshima and the nearby port city of Kure.[4] No doubt it was historical coincidence that Hiroshima also happened to be the prefecture that had sent the largest number of emigrants to the United States. The black hands of fate were about to tie those two circumstances together.

The minutes of the second meeting of the Atomic Bomb Target Committee, held May 10 and 11, 1945 in Oppenheimer's office, refer to the fact that "due to rivers," Hiroshima was "not a good incendiary target."[5] The rivers, that is, would reduce the effect of the conventional air raids using incendiary bombs that General Curtis LeMay had developed and used with devastating impact beginning with the massive air raid on Tokyo in March 1945. The fact that seven rivers flow through the heart of the city may have been the primary reason why Hiroshima had escaped conventional bombing.

The very fact that Hiroshima had been dismissed as a poor incendiary target, however, increased its suitability as a target for atomic bombing. The minutes of the initial meeting of the target committee on April 27, in which Major General Groves participated, reported that the 21st Bomber Command (in charge of conventional bombing) "has thirty-three primary targets on their target priority list" and "Hiroshima is the largest untouched target not on the . . . list. Consideration should be given to this city."[6]

At the second meeting, several additional factors emerged that argued in favor of Hiroshima as a target. The minutes include this note:

> Hiroshima—This is an important army depot and port of embarkation in the middle of an urban industrial area. It is a good radar target and it is of such size that a large part of the city could be extensively damaged. There are adjacent hills which are likely to produce a focusing effect which would considerably increase the blast damage.[7]

Hiroshima was subsequently placed near the top of the target list, second only to Kyoto. Each was classified an "AA target."

Specialists involved in planning the bombing continued to press for Kyoto as the target for the first bomb, despite the opposition of Secretary of War Stimson. The following classified message, dated July 21, was delivered to Stimson, who was attending the Potsdam Conference with President Truman: "All your local military advisors engaged in preparation definitely favor your pet city and would like to feel free to use it as first choice if those on the ride select it out of four possible spots in the light of local conditions at the time." The message was sent by George A. Harrison, deputy chairman of the Interim Committee on Atomic Bomb Affairs, who was acting as deputy in charge during the absence of the secretary of war.

Stimson's "pet city" was Kyoto, and the other three possible targets were Hiroshima, Kokura, and Niigata. Before he had left for Potsdam, Stimson had made it clear to Groves that he was absolutely opposed to bombing Kyoto. Faced with renewed pressure from his military advisors, Stimson issued definitive orders by way of classified messages through Harrison on July 21 and 23:

Aware of no factors to change my decision. On the contrary new factors here tend to confirm it.

Also give name of place or alternate places, always excluding the particular place against which I have decided. My decision has been confirmed by highest authority.

The highest authority, of course, was President Truman. The decision to use the atomic bomb rested with the president, but the selection of the target city was left in the hands of the secretary of war. Even so, Stimson went so far as to mention the authority of the president in order to protect Kyoto.

As a result, the only remaining AA target was Hiroshima. Later on July 23, Harrison replied to Stimson, listing Hiroshima, Kokura, and Niigata as the three targets.[8]

In the final bombing plan, Niigata was dropped from the list and Nagasaki was substituted. From the outset, Niigata had been classified only as a B target. It was also too far from Hiroshima, which was now the primary target. Nagasaki had been listed as a target for conventional bombing, and by the end of July it had been bombed five times, although it had not yet suffered a decisive blow. Some among the planners felt strongly that the effect of an atomic bomb on Nagasaki would be minimized by the hilliness of the city. But it was close to the Kokura arsenal, the secondary target, and in the end it was substituted for Niigata on the final list.

At 2:45 A.M. on August 6, 1945, the *Enola Gay* left Tinian. After circling to the left over Iwo Jima, the bomber chosen to carry the atomic bomb headed straight toward Hiroshima, following the invisible "Hirohito Highway" high over the Pacific Ocean to the mainland of Japan.

At 6:00 A.M., the night shift of the special intelligence squad of Nisei women that had been set up to intercept shortwave messages came off duty. Among them was Irene Ishigame, a member of the squad. She finished her breakfast at 6:30, and at 7:00 started back to her home in Nukushina.[9] Because her house was within walking distance of the headquarters, her life was saved. Others, whose homes were further away, stayed at the dormitory.

Irene reached home a little after 8:00 A.M., after about an hour's walk. Suddenly, there was a violent flash. Because her home was far from the hypocenter, no direct damage was caused. Irene at once started back toward Sentei to pick up her belongings, which she had left at the dormitory. She was only able to make it halfway. She realized then that it was more important to search for her younger sister Jane, who had been working in the Kowa Sewing Machine factory at the foot of the Yokogawa Bridge. In spite of Irene's efforts, however, Jane was not found until several days later, when a neighbor brought her home with injuries all over her body.

The mansion in Sentei was destroyed instantly by the bomb, and all the Japanese American women of the special intelligence squad who were on duty at that time are believed to have been killed. Still, the radio monitoring had to continue, so, on August 8, under the command of Lieutenant Colonel Kakuzo Ōya, several spare radios were dug out from where they had been kept underground. They were installed in a farm outbuilding on a hill in the town of Saijo, about twelve miles east of Hiroshima, and the interception of shortwave messages was resumed. Ōya himself had been at the headquarters of the Second Army at the time of the bomb and had suffered serious injuries to his head and neck. In his place, Major Tadashi Ishii, the chief of the Intelligence Section, whose wounds were minor, traveled back and forth between Saijo and Hiroshima to convey the monitored messages.

About ten members of the monitoring team, who had been off duty and therefore had survived the blast, were called on to resume work. Irene Ishigame was one of these. This time, no one on the team was allowed to return home. According to Ōya, the shortwave messages contained increasingly urgent and important information about the changing military situation, the Soviet entry into the war, and moves relating to the Potsdam declaration. All of this information had to be kept strictly secret.

With increasing frequency, the radio messages were predicting the defeat of Japan, which represented, to the Nisei women listening in, victory for their homeland. But even if some may have felt a certain pleasure in the news, there was no way they could express their feelings outwardly.

When the end of the war came, the members of the special intelligence squad shed tears with the soldiers at the farm in Saijo. But their feelings were far more complex. "I felt very uneasy," Irene admitted. "I thought the Americans would hate us. Come to think of it, we were traitors to the United States, weren't we?"

The atomic bomb destroyed half of the Army headquarters building in Hiroshima and, shortly afterwards, fire engulfed it and burned most of the rest of the building. With little loss of time, however, the headquarters was reestablished in a cave halfway up Mount Futaba, about a half mile northeast of the old site. In a meeting held there to review the situation, an officer expressed his suspicion that the blast might have been caused by an atomic bomb, although the message that was actually sent to imperial headquarters reported that "a special kind of high-yield bomb with powerful incendiary effect was dropped, destroying almost all of the city of Hiroshima."[10]

However, as "countermeasures to atomic bombing," the Hiroshima headquarters circulated the following advice:

1. If it is not a surprise attack like that on Hiroshima, large losses from the explosion of this type of bomb, particularly casualties to personnel, can be avoided by using air-raid shelters. The impact on the maintenance of ground forces will be minimal.
2. By covering oneself with something white, such as white cloth, incendiary effects of the flash can be avoided.[11]

It is likely that these countermeasures were forwarded to imperial headquarters. Subsequent directives issued by the Air Defense General Headquarters carried a note of reassurance, to the effect that the atomic bomb was "nothing to be afraid of."

The following advice appeared in Japanese newspapers: "If you are clad in white clothes that cover your skin completely, you will be protected from the heat rays. . . . Take shelter behind an object on the side away from the flash and you will not suffer a burn. . . . You will be safe in a covered shelter."[12]

The military never allowed the press to use the words "atomic bomb." While they sought to deny the power of the "new type of bomb," at the same time they stressed its inhumane character.

On the other hand, the official announcement by the United States government emphasized the vast power of the atomic bomb. This is how President Truman described to the world the dropping of the atomic bomb: "Sixteen hours ago, an American plane dropped a bomb over Hiroshima, an important military center of Japan. . . . It is an atomic bomb. It utilizes the fundamental power of the universe. . . . Let us make it clear. We will be completely destroying the war-making capabilities of Japan."[13]

American newspapers announced the news under such banner headlines as: "ATOMIC BOMB USED ON JAPS: Equivalent to 20,000 Tons of TNT" and "NEW WEAPON 2,000 TIMES MORE EXPLOSIVE THAN THE MOST POWERFUL BOMB IN THE WORLD." On the other hand, letter columns ran heated discussions about whether it was justifiable to use the atomic bomb on Japan, which was already so close to defeat. In the end, American public opinion seemed to accept the argument that use of the atomic bomb saved the lives of hundreds of thousands of American soldiers by bringing the war to an end.

While most Americans anticipated the imminent victory with great excitement and relief, some in the United States greeted the news of the atomic bombing of Hiroshima with the deepest anguish. They were the Issei and Nisei, Americans of Japanese descent, many of whom were still confined to the relocation centers. Of the various camps, the news created the greatest impact on Tule Lake. This was the isolated camp for those "disloyal elements" who had refused to pledge allegiance to the United States. Michi Weglyn writes:

> For the Issei and Nisei still trapped in Tule Lake, the atomic incineration of... kindred fellow humans in Hiroshima ushered in the final nightmare stage in a sequence of injustices which had issued forth from the order to evacuate. One-third of the segregant population were either natives of Hiroshima or had relatives living there, hundreds of them war-stranded Nisei.[14]

Marvin Opler, who was observing the situation at the camp on August 8 when the details of the news spread, wrote, "The atomic bombing of Hiroshima City had left the center stunned ... with a complicated series of reactions."[15]

In deep grief and sorrow, one third of the population of Tule Lake, the Hiroshima-born Issei, started to prepare a memorial service for their children and other relatives whom they assumed had been obliterated by the atomic bomb.

Notes

1. There were also rumors that Truman's cousin was living in Hiroshima; see Senshi Kenkyukai, ed., *Gembaku no ochita hi* (The Day the Atomic Bomb Fell) (Tokyo: Bungei shunjū, 1972), p. 166. The rumor that General MacArthur's mother was a Japanese born in Hiroshima appears in Tsuruji Matsuoka, *Hiroshima gembaku no shuki* (A Memoir of Hiroshima Atomic Bombing) (Private edition, 1960), p. 133.

2. Interview with Kakuzo Ōya, May 18, 1977, Tokyo.

3. Interview with Irene (Ishigame) Nakagawa, April 3, 1978, Los Angeles.

4. Hiroshima ken, ed. *Gembaku sanjū nen—Hiroshima ken no sengoshi* (Thirty Years After the Atomic Bombing: The Post-War History of Hiroshima Prefecture) (Hiroshima: Hiroshima Prefecture, 1975), p. 30.

5. Minutes of the Second Target Committee, May 5–11, 1945, in Manhattan Engineer District Records (hereinafter cited as *MEDR*), 5D–2. National Archives, Washington, D.C.

6. Minutes of the First Target Committee, April 27, 1945, *MEDR*, 5D–2.

7. Minutes of the Second Target Committee, *MEDR*.

8. Based on Otis Carey and Sojo Kitagaki, ed., *Bakugeki o manugareta Kyoto—Rekishi eno shōgen*. (How Kyoto Escaped the Atomic Bombing: A Testament to History) (Moonlight Series, Doshisha University Aamosuto Kan [Amherst House], 1975).

9. Based on interviews with Irene (Ishigame) Nakagawa and Kakuzo Ōya.

10. Kakuzo Ōya, "Report on the Situation Caused by the Atomic Bombing—Hiroshima, August 6, 1945" (Draft in Japanese). Provided to the author by Ōya.

11. *Ibid.*
12. *Yomiuri hōchi,* August 14, 1945.
13. *Public Papers of Harry S Truman,* August 6, 1945, pp. 196–200.
14. Michi Weglyn, *Years of Infamy* (Seattle: University of Washington Press, 1996), p. 339.
15. Cited in Weglyn, *Years of Infamy,* p. 400.

four

Heading Toward the Ruined City

Where the Tule Lake camp population was filled with deep grief for the relatives and next of kin who had become victims of the atomic bomb, anger and confusion were the dominant emotions at the detention camp in Santa Fe, New Mexico. Most of the detainees there were leaders of the Japanese community who had been rounded up shortly after the attack on Pearl Harbor. They had earlier been confined in various other camps, but beginning in June 1943, most had been transferred to the Santa Fe camp. In the last year of the war, this camp contained in microcosm the whole Japanese community in the United States.[1]

The Los Alamos atomic bomb production facility was situated some twenty miles northwest of the camp, but none of the detainees knew of the project, which had been cloaked in absolute secrecy. The first test detonation of the atomic bomb took place on July 16 in Alamogordo, New Mexico, about two hundred miles due south of Santa Fe. Major General Groves reported that the flash of the test explosion was seen from Santa Fe, but there is no record that any of the detainees saw it.[2]

The vast majority of the detainees at Santa Fe were Issei. Because of their position as leaders in the Japanese community, they had been treated as likely subversives from the time Japan launched its war on the United States. The detainees had been separated from their wives and children, and transferred from one camp to another. Many continued to harbor the hope that Japan would win the war.

Late in 1944, the government transferred a large number of "radicals" (young Kibei Nisei who called themselves the "Loyal Youth Group" and pledged allegiance to Japan) from the Tule Lake camp to Santa Fe. This move intensified antagonism towards the United States among the Santa Fe detainees and threatened to overwhelm the remaining modicum of rational thinking in the camp.

When the news of Japan's defeat arrived, few of the detainees believed it. On the morning of August 10 the first reports of Japan's impending surrender were broadcast on the camp's public address system, but the detainees decided that the news was a false rumor aimed at disguising the defeat of the United States, since it was unthinkable that Japan could be defeated. Then, all through the night of the 13th into the morning of the 14th, jubilant Santa Fe residents could be heard celebrating the American victory. Still the detainees did not believe the news of defeat. The atmosphere in the camp made it impossible to broadcast President Truman's announcement of the end of the war, or to report it in the Japanese-language newspaper circulated among the detainees.

Soon, the camp was divided into a "Japan won" group and an "unpatriotic" group. The latter consisted of intellectuals who accepted the fact of Japan's surrender, but they represented perhaps one-tenth of the detainees. The remaining ninety percent insisted that Japan had won and started to spread rumors within the camp, including stories that Pearl Harbor and Honolulu had been destroyed by a "double atomic bomb," that Japan was disarming the United States, and that the Panama Canal had been blown up.

On the evening of September 14, a memorial service for the victims of the war was held at the camp hall, sponsored by the Buddhist Federation. After sutras were chanted for the souls of both Japanese and American war dead, the Rev. Enryo Shigetoh of the Fresno Temple of Nishi-Hongwanji delivered a sermon. He poignantly protested the illegality and unfairness of the atomic bomb, and then continued: "The whole nation of Japan was outraged at this extremely inhuman act of the United States. As a result, the Japanese forces delivered a deadly blow to the American expeditionary troops, and within three days, forced the United States to sign an article of surrender to Japan."[3]

How these people wished for Japan's victory! Those in the "Japan won" group not only hoped for it, but refused for weeks and even months to believe the reports of Japan's shattering defeat. On the morning of November 23, more than two thousand detainees, a majority of those at the Santa Fe camp, voluntarily left for Japan. A large number of Kibei Nisei from Tule Lake who had refused to take the loyalty oath joined the Issei on the return voyage, even though it meant renouncing their American citizenship. Japan remained their homeland, and some of them arrived there at year's end still doubting the American claims of victory. It was only after personally witnessing the ruins of Tokyo and Hiroshima that they finally accepted the reality of Japan's defeat.

The first foreign reporter to reach Hiroshima was an Australian journalist named Wilfred Burchett. Burchett was in Okinawa when he heard the news of the atomic bombing, and decided to go directly to Hiroshima as soon as he reached Japan. On September 2, while the attention of most foreign correspondents was focused on the surrender ceremony in Tokyo Bay, Burchett headed for Hiroshima by train.

What he saw in Hiroshima the next day were ruins that extended as far as the eye could see. Everything was permeated with the smell of death. "I had the feeling of having been transplanted to some death-stricken alien planet," he wrote. "There was devastation and desolation, and nothing else."[4] He witnessed people who were still dying as a result of the radiation, even though it was almost a month after the bombing.

The dispatch he sent from Hiroshima to the *Daily Express* of London begins: "In Hiroshima, thirty days after the first atomic bomb destroyed the city and shook the world, people are still dying, mysteriously and horribly—people who were uninjured in the cataclysm—from an unknown something which I can only describe as the atomic plague."[5]

Soon after Burchett handed this article to a member of the Domei News Service, U.S. correspondents who had come directly to Japan from Washington arrived in Hiroshima. According to Burchett, "They saw physical wreckage only." These "top-flight correspondents," he wrote, did not even try to see the human misery brought about by the atomic bomb.

Several days later, Burchett arrived back in Tokyo and barely made it to a U.S. Army press conference at the Imperial Hotel. He learned that the sole purpose of the press conference was to deny that people were still dying from what he had called the "atomic plague." "An American scientist in Brigadier General's uniform," he wrote, referring to Thomas F. Farrell, leader of the Manhattan Project Investigating Group, "was explaining that there could be no question of atomic radiation because it had been ensured that the bomb exploded at a high enough altitude to obviate any risk of radiation in the soil."[6]

There was one major weakness in the testimony of those who denied that any deaths had occurred due to radiation: They had not yet visited Hiroshima. Burchett disputed their facts, referring to what he had seen with his own eyes, but the U.S. Army refused to hear him out. Furthermore, it designated Hiroshima "off limits" to all foreign correspondents. On September 19, a press code that included a prohibition on news about the atomic bomb was imposed on all Japanese newspapers, broadcasters, and publishers by General Douglas MacArthur's headquarters.

The dropping of two different types of atomic bombs on Hiroshima and Nagasaki resulted, in one sense, in a test of the effect of the bombs on civilian populations. An immediate effort was mounted to investigate the results. One team, the Manhattan Project Investigating Group, left Hamilton Air Base in California as early as the morning of August 13, bound for a base in the Marianas, where they remained on stand-by after August 15. Additional survey teams were organized by both the Army and Navy, as well as by the United States Strategic Bombing Survey Team (USSBS).

Authorization for the surveys had to be obtained from the U.S. Pacific Command under General MacArthur, who was responsible for the occupation of Japan. On August 28, MacArthur's medical advisor, Colonel Ashley W. Oughterson of the Medical Corps, drew up a diagram of the investigation while on board the *USS General Sturgis*, which was carrying officers of the occupation force from Manila to Yokosuka. This memorandum, entitled "Study of Casualty Producing Effects of Atomic Bombs," began as follows:

> A study of the effects of the two atomic bombs used in Japan is of vital importance to our country. This unique opportunity may not again be offered until another world war. Plans for recording all of the available data, therefore, should receive first priority. A study of the casualty producing effects of these bombs is a function of the Medical Department and this memorandum is prepared as a brief outline for such a study.[7]

Advance estimates of casualties from the atomic bomb had been relatively low, and Oughterson's memorandum, which is interesting in many ways, reflects this consistent tendency to understate the number of victims. His memorandum declared, "The total number of casualties reported at Hiroshima is approximately 160,000 of whom 8,000 are dead." Oppenheimer was reportedly shocked to learn that the actual number of casualties in Hiroshima greatly exceeded the twenty thousand that had been forecast. Even as the investigation progressed, survey groups continued to underestimate the number of those killed instantaneously.

The memorandum suggests that the investigation was planned on a large scale: "All living casualties should be identified by number for location (at the time of bombing) on the map and an exact description of the case kept in a cross index file." At this point, there were as many as 150,000 survivors, far beyond the scope of the initial survey. The effort to ascertain the overall condition of survivors would later be taken over by the Atomic Bomb Casualties Commission (ABCC).

Even though the investigation was nominally to be conducted for medical purposes, its real subject was the "study of casualty producing effects of atomic bombs," and its outline contained detailed instructions on how to examine each casualty. It is noteworthy that special attention was paid to the existence or nonexistence of shelters and the nature of shielding between the blast center and each casualty. Of course, this was essential to estimating the amount of radiation to which each casualty was exposed. But it is likely that the investigation also sought ways to minimize the effects of future atomic bombs. "The position or protection of all casualties should be determined since this may be a determining factor in blast effects and burns. (Standing, sitting, prone, indoors, outdoors, in shelters, trenches or behind walls, etc.)"

In short, the investigation was clearly aimed at observing the effects of the atomic bomb rather than offering necessary treatment for casualties. The rationale was that of an aggressor: investigation and research, but no treatment. This later became the policy of the ABCC, although in later years, the results of some of the research were made available to local medical practitioners who were treating survivors.

Near its end, Oughterson's memorandum gives an ominous warning to all who would take part in the investigation: "It should be emphasized that since the effects of atomic bombs are unknown, the data should be collected by investigators who are alert to the possibility of death and injury due to as yet unknown causes." There is a saying, "Go for wool and come home shorn." The atomic bombs dropped for the first time in history made the citizens of Hiroshima and Nagasaki guinea pigs of a unique experiment. But the new weapon contained unknown dangers, and it was possible that those who carried out studies on these guinea pigs would themselves end up as test samples.

Oughterson's memorandum was submitted to Brigadier General Guy B. Denit, surgeon general at the general headquarters of the U.S. Pacific Command, and it was approved by him at once. It would still be some time, however, before the full-scale investigation could be launched.

The investigative team slated to enter Hiroshima first was heading toward Japan at that time aboard the *USS Sturgeon*. Among the passengers was Colonel Crawford F. Sams, chief of the public health section of the U.S. Pacific Command, who was to perform the same function in the General Headquarters (GHQ) established for the occupation. As he expressed it, the *Sturgeon* passenger list "reads like a *Who's Who* of the Army, Navy, and Army Air Force of the Pacific War."[8]

The Manhattan Project Investigating Group was led by Brigadier General Thomas F. Farrell, who joined the party while they were standing by at the base in the Marianas. According to Aberill Liebow's "Encounter with Disaster," "their mission was to conduct a brief and preliminary survey on the effects of the atomic bomb and report the results to Washington directly. Their major role was to measure the residual radiation to guarantee security for the Occupation Army."[9] In the official report of the investigation team, submitted on June 30, 1946, its aim was: "(1) To make certain that no unusual hazards were present in the bombed cities. (2) To secure all possible information concerning the effects of the bombs, both usual and unusual, and particularly with regard to radioactive effects, if any, on the targets or elsewhere."[10]

The chief of the medical department of the Manhattan Project group was Colonel Stafford L. Warren, who was a professor in the Rochester University Department of Radiology and had been acquainted with Oughterson for many years prior to his military service. Warren arrived in Japan on September 7. Oughterson had already met with the Manhattan Project group, on September 4, two days after Japan's formal surrender on board the *USS Missouri*, and at that time an agreement was reached that all results of the investigations would be unified and compiled in joint medical reports.

But would it be possible to send an American survey group safely into an area where people were still suffering from the injuries caused by the atomic bomb? The area that included Hiroshima and Kure was not within the jurisdiction of American occupation forces, and several weeks would elapse before the British Commonwealth troops assigned to the region arrived in Japan. What would happen to a small group of Americans in that environment? Colonel Sams, chief of the public health section of the GHQ, recalled more than thirty years later that he seriously feared that "the people of Hiroshima would hang us from the trees."[11] Colonel Warren also related, "I did not expect to come back alive, for I thought the people of Hiroshima must be burning with anger."[12]

An unexpected opportunity presented itself. Just after landing in Japan, the GHQ set up a temporary office in the customs house in Yokohama, where Sams occupied one corner. Dr. Marcel Junnot and Margarita Strala of the International Red Cross came to him, reported on conditions in Hiroshima, and requested Sams to send medical supplies that were in seriously short supply because of the destruction of the hospitals. Sams hit upon the idea of taking advantage of this appeal to provide cover for the investigating group to enter Hiroshima. Thus, the Manhattan Project group arrived in Hiroshima on September 8, accompanying seven Army DC-6 cargo planes filled with medical supplies.

Actually, the people of Hiroshima had experienced such terror, as if the heavens had fallen upon them, and they were so completely dazed by their defeat in war, that they had lost the energy necessary to even shake their fists at a group of American investigators. Every member of the subsequent survey groups felt the same initial anxiety, followed by relief as they realized that the people of Hiroshima did not display any hostility or hatred toward the arriving Americans.

Sams also recalled that, throughout the years of the occupation, not one of the hibakusha declared to him that he or she was of American nationality.[13]

The American survey team had yet to discover that there were American POWs among the victims of the Hiroshima bombing. The POWs had been confined in several places in the city, and some were killed instantly by the bomb. Others who survived were dragged into the streets, to be stoned and tortured to death by enraged Hiroshima citizens. This grim episode—including the unfounded rumor that there were two women among the POWs—circulated among survivors, but seems never to have reached the survey team. Twenty-three dog tags from American POWs, marked "killed by the atomic bomb" were placed in the safe of the Hiroshima military police. A report later submitted by the Japanese government to the occupation authorities disclosed twenty names (including one who remained unidentified) as "those who died in Hiroshima by the atomic bombing."[14] Only a very few members of the Japanese Army knew the much grimmer fact that eight of the dog tags belonged to American POWs who were victims of vivisection experiments conducted until May 1945 at the medical school of the Kyushu Imperial University. The Pentagon has never confirmed the names or the causes of death of these eight victims or the other American POWs who were killed by the bomb.[15]

How long the Manhattan Project group stayed in Hiroshima is the subject of some dispute. According to Liebow's "Hiroshima Medical Diary," the group returned to Tokyo after ten days in Hiroshima, having concluded that there was minimal residual radioactivity and the city was safe for occupation forces.[16] However, a report compiled by Hiroshima Prefecture thirty years after the atomic bombing states that the group returned to Tokyo on the same day it entered Hiroshima.[17] It appears, in fact, that a preliminary survey was conducted for two days, September 8 and 9, followed by a full-scale investigation for an additional four days. Brigadier General Farrell returned to the United States, carrying all the data with him.[18]

While the stated purpose of this survey group was to investigate residual radiation, Colonel Sams, who accompanied the group to Hiroshima, observed that they had a much larger purpose from a political standpoint—to inform the people of the world of how devastating an atomic bomb could be.[19] The United States at that point had exclusive possession of the weapon that could wipe out a whole city with a single shot, and it would naturally wish to impress future enemies with the enormous power of nuclear weapons.

It is interesting that the danger of residual radiation was minimized in the survey report. The first major conclusion of the report stated that "the residual radiation after the explosion is not existent in a hazardous volume."[20] An entirely different opinion was proffered a year and a half later by Stafford Warren, the chief of the medical division of the survey group. In February 1947, as he assumed a professorship of biology in the medical school at UCLA as a top scientist in the area of nuclear medicine, he is reported to have said, "If an atomic bomb is dropped on Los Angeles, the whole city will be barren for fifty to one hundred years to come." Warren made this observation just after he had been involved in Operation Baker, a series of atomic test explosions underwater near Bikini Island in the Pacific. He also predicted that the next world war would bring about the self-destruction of the human race, stating that "while all scientific knowledge is brought to bear on measures to protect against the atomic bomb, it is impossible within the limits of current science."[21]

Colonel Sams, however, reached a different conclusion. On the basis of interviews he conducted in Hiroshima, he estimated the number killed instantaneously by the atomic bomb at two thousand.[22] Another 20,000 people, Sams calculated, had been buried under fallen buildings or burned to death within twenty-four hours of the bombing.

Including those who subsequently died due to radiation and burns, Sams concluded that the total number who perished in Hiroshima as a result of the atomic bomb did not exceed 67,000. This figure, he explained, included "those among the people who were staying in Hiroshima City at the time of bombing, those who died within six months after the bombing, and those who entered the city immediately afterward." Sams also insisted that if appropriate medical care had been supplied to the casualties, fifty percent of those who died could have been saved. His conclusion? "An atomic bomb is a poor massacre weapon. It is a damned poor killer." Sams also expressed confidence that people can protect themselves from atomic bombing with appropriate shelters, and that ninety percent of the population could survive a nuclear war.[23]

The survey plan Oughterson drafted on board the *USS General Sturgis* at the end of August 1945 read in part: "It is hoped that some post-mortem examinations may have been done by the Japanese and that these records may be amplified by early interrogation of the Japanese pathologists."[24]

A variety of Japanese survey groups were, in fact, organized and dispatched to Hiroshima by the Army, Navy, various universities, and the prime minister's cabinet. Some were organized as early as the day after the bombing. The best-known of these was the Army survey group led by Lieutenant General Seizō Arisue, among whose members was Yoshio Nishina of the Institute of Physical and Chemical Research. When he saw the damage that had been done to the city, Nishina immediately understood that the devastation had been caused by an atomic bomb. Before the end of the war on August 15, the priority for all of the Japanese groups was to determine if the explosion was caused by an atomic bomb and to develop countermeasures based on the condition of the casualties.

After the war ended, the emphasis of the Japanese survey groups shifted to treatment and relief for the survivors. A full-scale investigation by authorities in various fields began on August 30, with the arrival in Hiroshima of a research and medical group including Professor Masao Tsuzuki and Assistant Professor Jinsuke Miyake, both of Tokyo Imperial University, and Asao Sugimoto of the Institute of Physical and Chemical Research.[25]

The aim of the American groups, as we have seen, was to gather strategic information. The three groups sent by the United States—GHQ's group of physicians (twenty-three members) led by Colonel Oughterson, the Manhattan Project group under Brigadier General Farrell (thirty members), and a U.S. Navy group of medical specialists (fourteen members) led by Colonel Shields Warren, chief of the Navy Medical Corps—were brought together into a unified effort. With the assistance of ninety Japanese specialists led by Dr. Tsuzuki, the "Joint Commission for Investigating the Effects of the Atomic Bomb in Japan" was established. The commission's operations were delayed at the outset because of a typhoon. Survey activities began October 14, from a headquarters set up in the Ujina Army Hospital.

Although it was called a "joint commission," the initiative was always taken by the American victors, with their priority placed on data for strategic purposes. The data that were collected, together with information obtained earlier by the Japanese groups, were all delivered to the Army Pathological Institute in Washington by the American survey groups around the end of 1945. Nineteen reels of a movie, "The Effects of the Atomic Bomb: Hiroshima and Nagasaki," that had been shot by a crew from Nihon Eigasha (Japan Movie Company) were also confiscated and sent to Washington, along with the negatives and the outtakes. All data on the damage inflicted by the atomic bomb became the exclusive property of the United States. This material, twenty thousand items in total, remained unavailable to the Japanese until twenty-eight years later, when it was returned to Japan.

While the occupation press code initially prohibited any reference to the effects of the bomb in Japan, the Japanese community in the United States was presented with a grim account of how the victors were treating their atomic-bomb victims. This appeared in the *Rafu Shimpō*, a newspaper written in Japanese and English and published in Los Angeles. In an article headlined "Atomic Bomb Corpse Closely Studied by Army Department in Washington," dated July 1, 1946, readers were informed that "[p]arts of the dead body of Midori Naka, thirty-five years old, who was a beauty among Japanese actresses and one of the casualties of the atomic bomb dropped on Hiroshima City last summer, are said to be undergoing close study at the Pathological Institute of the Army in Washington, D.C., for the purpose of investigating what sorts of damage an atomic bomb can cause to human cellular tissue."[26]

At the time the atomic bomb was dropped, Naka was in the dormitory kitchen of the office of the Japan Traveling Theater Federation in Hiroshima, where she was performing. She suffered only a small external wound, but after returning to Tokyo on August 9 she was admitted to Dr. Tsuzuki's surgery at Tokyo Imperial University, where she died on August 16. In the "Interim Report on the Atomic Bombing Survey by the Survey Group of the Institute of Physical and Chemical Research," there are such notations as "(August) 29: No strong radioactivity observed in the thighbone of Midori Naka" and "(September) 1: Measurement of Midori Naka's skull."[27]

After the post-mortem examination at Dr. Tsuzuki's surgery the American investigating group confiscated Naka's remains and all records pertaining to her. "A part of her remains was sent from Tokyo to Washington," the *Rafu Shimpō* reported. "Detailed, post-mortem examinations already have been conducted on 175 casualties," the article continued, "and it was clarified that there is some relationship between the houses where the victims were at the time of the bombing and the part of their body that experienced the fatal wound. Based on the results of these examinations, a study of buildings to prevent damages from atomic bombs will be carried out."[28] It is a strategic principle that a party that develops weapons is required to develop countermeasures to protect against their destructive force. The investigation of the effects of the atomic bombs dropped on Hiroshima and Nagasaki was a step toward preparation for future war.

The article on Midori Naka noted, "The investigation of the damage to human bodies is aimed at the development of the remedies." Whatever remedies the Army's Pathological Institute developed, they would not be extended to the hibakusha in Japan. The United States was primarily interested in the study of long-term effects of an atomic bomb upon human beings and not in providing immediate remedies to the survivors of the Hiroshima and Nagasaki bombs.

On May 15, 1946, Colonel Oughterson sent a short memorandum to the surgeon general of the U.S. Army, entitled "The Need for Continued Study of the Atomic Bomb Casualties in Japan." The third item in the memorandum states: "There exists a unique opportunity to make long-term studies on the effect of an instantaneous dose of multimillion-volt X-rays on growth phenomena. There is every reason to believe that the carcinogenic effects of these X-rays as well as the flash burns are of the greatest importance."[29]

On November 26 of the same year, President Truman signed an executive order directing the National Academy of Science to conduct long-term research on injuries caused by the atomic bomb.[30] The principle was by now well established that treatment for hibakusha would not be considered the responsibility of the American government. The Atomic Bomb Casualties Commission (ABCC) was therefore established as an institution for research only.

Notes

1. Dōjun Ochi, ed., *Minami-Kashu nihonjinshi* (History of the Japanese people in Southern California) (Los Angeles: Nanka Nikkeijin Shōgyō Kaigisho, 1957), vol. 2, p. 266.

2. Leslie Groves's report in Manhattan Engineer District Records, in Martin J. Sherwin, *A World Destroyed: The Atomic Bomb and the Grand Alliance* (New York: Knopf, 1975), Appendix P, p. 308.

3. Ochi, *Minami-Kashu nihonjinshi*, pp. 303–308.

4. Wilfred G. Burchett, *Passport: An Autobiography* (Australia: Thomas Nelson, 1970), p. 167.

5. *Daily Express* (London), September 5, 1945.

6. Burchett, *Passport*, pp. 175–176. It should be noted that American investigators were, in fact, aware that there was a possibility that Hiroshima was contaminated by radioactivity. One of the assignments of the investigating group was to determine the levels of residual radioactivity in Hiroshima and Nagasaki, to insure the safety of arriving Allied armed forces. For a summary of these surveys, see Barton C. Hacker, *The Dragon Tail: Radiation Safety in the Manhattan Project* (Berkeley: University of California Press, 1987).

7. This and excerpts quoted below are found in Aberill Liebow, "Encounter with Disaster: A Medical Diary of Hiroshima, 1945," *Yale Journal of Biology and Medicine*, Vol. 38, October 1965, pp. 82–83.

8. Crawford F. Sams, *Medic* (manuscript), p. 339.

9. Liebow, "Encounter with Disaster," pp. 84–85.

10. United States Department of War, Plans & Operations Division, "The Atomic Bombing of Hiroshima and Nagasaki," File No. P & O 000.9 (Sec. V), cases 51–60, National Archives, Washington, D.C.

11. Interviews with Crawford F. Sams, October 24, 1975, and April 11, 1977, Atherton, CA. Also Sams, *Medic*, p. 361.

12. Interview with Stafford Warren, April 25, 1978, UCLA.

13. Interviews with Sams.

14. "Location of Remains of Allied Personnel" prepared by Central Liaison Office, Tokyo (CLO #1040–5,3), and subsequently submitted on March 5, 1946, to GHQ. Declassified by the Ministry of Foreign Affairs in June 1977. (MFA Microfilm, Reel A–0118, Flash 134–5).

15. Two female POWs are depicted in the well-known murals of the Hiroshima bombing executed by the Japanese artist couple, Iri and Toshi Maruki. See John W. Dower and John Junkerman, eds., *The Hiroshima Murals: The Art of Iri Maruki and Toshi Maruki*, particularly, "The Death of the American POWs" (Tokyo and New York: Kodansha International, 1985).

On the dog tags of POWs, see *Gembaku sensaishi* (Record of the Atomic Bombing Suffering in Hiroshima) (Hiroshima: Hiroshima shi [City], 1971), Vol. 1, pp. 175–180.

On vivisection, see Kyodo News Service, July 24, 1978; and Fuyuko Kamisaka, *Seitai kaibo: Kyūshū daigaku igakubu jiken* (Vivisection: The Case of the Kyushu University Medical School) (Tokyo: Mainichi shimbunsha, 1979).

16. Liebow, "Encounter with Disaster," p. 84.

17. Hiroshima ken, ed., *Gembaku sanjū nen—Hiroshima ken no sengoshi* (Thirty Years After the Atomic Bombing: The Post-War History of Hiroshima Prefecture) (Hiroshima: Hiroshima Prefecture, 1975), p. 75.

18. U.S. Department of War, Plans & Operations Division, "The Atomic Bombing of Hiroshima and Nagasaki," p. 4.

19. Interviews with Sams.

20. U.S. Department of War, Plans & Operations Division, "The Atomic Bombing of Hiroshima and Nagasaki," p. 4.

21. *Rafu Shimpō* (A Japanese vernacular newspaper in Los Angeles), February 5, 1947 (retranslated from the Japanese).

22. Interviews with Sams. In the April 1977 interview, Sams reaffirmed his casualty estimates.

23. The most comprehensive study of the casualties of Hiroshima and Nagasaki can be found in The Committee for the Compilation of Materials on Damage Caused by the Atomic Bombs in Hiroshima and Nagasaki, *Hiroshima and Nagasaki: The Physical, Medical, and Social Effects of the Atomic Bombings* (New York: Basic Books, 1981). This study estimated that 140,000 people were killed in Hiroshima and 70,000 in Nagasaki.

24. In Liebow, "Encounter with Disaster," p. 83.

25. Hiroshima Ken, ed., *Gembaku sanjū nen—Hiroshima ken no sengoshi* (Thirty Years After the Atomic Bombing: The Post-War History of Hiroshima Prefecture) (Hiroshima: Hiroshima Prefecture, 1975), p. 74.

26. *Rafu Shimpō*, July 1, 1946.

27. Hiroshima Ken, *Hiroshima kenshi* (History of Hiroshima Prefecture), *Gembaku shiryō hen* (Materials on the Atomic Bomb) (Hiroshima: Hiroshima ken, 1972), p. 490.

28. *Rafu Shimpō*, July 1, 1946.

29. U.S. Department of War, Plans & Operations Division, "Atomic Bomb Casualty Studies," Tab B.

30. *Ibid.*, Tab F.

five

Nisei Coming, Nisei Going Home

As the Allied Occupation came into full swing, occupation forces began to appear throughout Japan, but few American military personnel were seen in Hiroshima except for occasional sightseers and those on survey assignments. Possibly because the United States wished to avoid the area, British Commonwealth forces occupied the Chugoku region, which includes Hiroshima.

A Japanese-American hibakusha, who later returned to the United States, vividly recalled her first encounter with American soldiers riding in jeeps. "I was terrified," she remembered. "I was trying to be truly Japanese, so I pretended not to understand English when they started talking to me." Frightened though she was, she felt no hatred toward Americans for having dropped the atomic bomb. She did not know why this was so. She had been raised in the United States until she was thirteen, but after spending the war years in Japan, American soldiers looked like aliens to her.[1]

Of course, Japanese Americans living in Japan did not feel the same fear of Nisei soldiers who visited Hiroshima, even though they wore the same uniform as the other GIs. Most of these Nisei soldiers had relatives in Hiroshima or were connected to the city in some other way. These Nisei had joined the military as draftees or volunteers, after having faced the psychological agony of taking the loyalty oath in the relocation camps.

A large number of Nisei soldiers had been sent to the front, most famously in the 442nd Regimental Combat Team of the Army's 100th Battalion, which saw heavy action in Europe. But the U.S. military did not assign Nisei soldiers to the battlefield in the Pacific theater, because they would be at risk from friendly fire since they were indistinguishable from the enemy. Instead, the Nisei were deployed in support roles, often in the Military Intelligence Service where they worked in translation, interception of radio communications, interrogation of Japanese POWs, and the production of psychological warfare leaflets and broadcasts aimed at the Japanese.[2] Japanese language specialists, who were mainly Nisei, were seen as the U.S. military's secret weapon in the Pacific. General Charles Willoughby, chief of intelligence under General MacArthur, commented that the contributions of Nisei soldiers shortened the duration of the war by as much as two years.[3] After the occupation got under way, the role of Nisei military personnel fluent in Japanese became even more critical. They were now part of a victorious army occupying the homeland of their fathers and mothers.

Many Nisei GIs had families and relatives in Hiroshima who were killed by the atomic bomb, but others were lucky enough to be able to reunite with parents, siblings, and relatives from whom they had long been separated. They were able to pass on news from the States and share supplies with loved ones who were facing shortages of food and daily necessities.

Sadako Obata was one of those who benefited from such a visit. As described earlier, she was five months pregnant when she was injured by the bomb. After taking refuge at a relative's house in the countryside, she delivered a baby girl in December. Because her husband died from acute radiation sickness, Sadako went through the hardships of childbirth on her own. In the midst of this predicament, her brother—a Nisei soldier stationed in Japan—paid a visit. They had not seen each other for seven years, since Sadako left for Japan in 1939 to visit her ailing grandmother. She recalled the emotional reunion: "I was very happy to see him again because he was the only one in Japan I could call my family," she said. "After that, people around me started to treat me much better because my brother brought food supplies and various other necessities with him."[4] The years of self-denial were over, years of suppressing the American within her out of fear that she would be treated as an enemy.

Nisei living in Japan gradually became aware that they were citizens of the country that had won the war. At the same time, they asked themselves, "Can we return to the United States? And if so, when?" They felt only uncertainty. The brothers, relatives, and friends who came to Japan as conquerors were all they had to rely on.

Judy Enseki, who had come to Japan on an exchange ship during the war, described her life in wartime Japan as "a fish out of water." Like many others, she was able to obtain a job with the occupation forces through personal connections—a cousin who worked as a civilian for the occupation forces helped her land a job at GHQ. There were many opportunities available as interpreters for older Nisei who still remembered English. Enseki left Hiroshima on a special occupation train for Tokyo.[5]

Not all of the Nisei who arrived in Japan came as victorious soldiers. Others had chosen to abandon their U.S. citizenship, little realizing that by doing so they had thrown in their lot with the losers of the war. These were mostly Kibei Nisei, who had been raised in Japan and then returned to the United States before the war broke out. They had chosen to abandon their U.S. citizenship and return to Japan while they were still confined in relocation camps. It is said that leaders among the Issei who maintained an unfaltering allegiance to Japan exerted a strong influence over the Kibei Nisei in the camps.

On November 24, 1945, the first group of repatriates left Tule Lake to embark on the initial postwar exchange ship, which sailed from Seattle the next day. After disembarkation at Uraga, a port near Tokyo, on December 6, they were temporarily housed at an Imperial Japanese Navy dormitory in Kamoi, a mile and a half inland. The shock of recognition of the reality of a defeated Japan destroyed whatever hopeful illusions they still carried with them.

One of the returnees wrote from Kamoi to his brother at Tule Lake, "Although we were told Japan was defeated, I held onto a faint hope. I could not accept what I had read in the American newspapers and *Life* magazine. Wasn't that propaganda? That's what I felt until we arrived here. I have now seen the devastation of Japan, and I am forced to admit that the photos we saw in *Life* magazine were not a deception."

He continued, "Even the tough men from the activist group at Tule Lake were disheartened at the reality of Japan. The returnees who set foot in Japan after requesting repatriation have now, without exception, admitted they should have remained in the United States. Please pass the word on to everyone there and ask them to reconsider."[6] Even taking Japan's defeat as a given, no one anticipated such devastation. How much greater the shock must have been for those who returned believing Japan had won the war. Another repatriate reported, "I only realized Japan had been defeated when I landed at Uraga and saw American soldiers on sentry duty."[7]

Irene Ishigame told this story about her father, one of the Nisei who returned to Japan:

Mr. A. came back to Hiroshima from the United States in December 1945 and told me that my father had returned by the same ship. I assumed that he must have been reluctant to get in touch with us because he had returned alone, leaving the rest of the family behind. I wondered about him, knowing that most likely he had no money, no extra clothes and no way of obtaining food, which was rationed. My grandfather asked me to bring my father back to Hiroshima, so although I had never been to Tokyo before, I got on a train packed with passengers. I was carrying a supply of rice and food for three days. It took twenty-four hours to get to Tokyo Station, where I changed to the Keihin line. Somehow I managed to find my way. I was able to track down the Kamoi repatriates dormitory, but when I asked about my father, I was told that he had gone out a little while earlier.

I put down my bags and ran out immediately to search for him. As I crossed the overpass on the way to the Uraga station, I caught sight of a miserable man in overalls walking toward me, looking like a straggler from a defeated army. Blood ties make for strange reactions. I stopped in my tracks to stare at this pathetic figure, and I realized it was my father. I squatted down and cried. He told me that he wept every night with regret, and then we both cried.

The Buddhist priests at Tule Lake filled my father with dubious information, telling him that he would be given a regular allowance if he returned to Japan. He seemed to have believed that Japan had actually won the war. But I think he was put on the ship back to Japan as a "traitor" because he had refused to pledge allegiance to the U.S. government.

He told me he could not contact the rest of the family before he left because they had been assigned to a different camp. I talked to him about the atomic bomb, and told him not to be shocked when he saw how Hiroshima had been burned to the ground. I reassured him that our house was still intact, and that my younger sister had survived the bomb and was recuperating. I begged him to come back to Hiroshima with me, but he was adamant about not returning.

There was no point in arguing, so I took out the food I had brought. He devoured three days' worth of food in one sitting. It's not surprising, since my father was being fed a diet of stewed rice and radish leaves. I returned to Hiroshima alone, leaving him with money and rice from my knapsack.

Some time later, my father came back to Hiroshima with three other stubborn friends. In February of the following year, my mother and four brothers and sisters returned to Japan together with my uncle's family in a group of more than ten. My parents abandoned all of their property, including their land in Fresno. Luckily, my grandfather owned a good-sized piece of property, but the house was crowded and the family got no rations because they were mostly American citizens. My parents never went back to the United States. They worked as cooks at a hospital and farmed to support themselves and five children. I have never known anyone who experienced a harder life than my parents.[8]

One observer who reported on the returning Nisei at the time described their situation in this memorable passage:

"Melancholic rain" they call it here, and surely the rainy season in Japan casts a special gloom over the people who used to live under the dry blue skies of California. Those Americans, who had forgotten what umbrellas and raincoats are for, now walk in the rain without them, like wet rats. That is a common sight in Yokohama these days. Those who look like wet, grimy American beggars are mostly young Kibei Nisei repatriates who gave up their citizenship in the U.S.[9]

The Nisei who spoke English sometimes found opportunities with the occupation forces; however, the observer reported,

There is discrimination at job sites according to their citizenship status. To begin with, wages are discriminatory. Nisei with U.S. citizenship are paid around 3,500 yen ($10) monthly, while those who have renounced their citizenship are paid only 500 yen ($1.50) for the same work. The ex-citizens are not entitled to use the exclusive trains of the occupation forces, nor are they allowed to eat the nutritious lunches and dinners available in the occupation dining halls. . . .

The ex-citizens don't bathe for days on end. Desperation drives these men in dirty clothes to gamble. Playing poker, blackjack, and rummy night after night in cold rooms deep in distress, sometimes chairs will fly and blood will flow from cuts opened by broken beer bottles. These are men who returned to Japan expecting to share the glory of a victorious nation—not, as it has happened, to work building a new Japan. None are more pathetic than these repatriated Nisei.[10]

In contrast to the repatriates who had gambled on the wrong side, the Nisei who suffered through the war in Japan saw the conclusion of the war as the end of their struggle and an opportunity to return to their homeland. After all, they were American citizens, even if they had been living lives thoroughly integrated into Japanese society.

According to an estimate made by the U.S. Consulate in Yokohama, there were approximately fifteen thousand Nisei residing in Japan as of the end of the war, of whom ten thousand were considered eligible to return to the United States.[11]

U.S. citizenship was automatically revoked for any Nisei who had served in the Japanese military. The same was true for those who had voted in any Japanese election. When the first general election was held after the war in April 1946, General MacArthur strongly encouraged every adult to vote, in support of the effort to democratize Japan. Ironically, many Nisei in Japan unwittingly voided their U.S. citizenship by joining this effort and casting their votes. Women were granted suffrage for the first time in the nation's history, and many Nisei women exercised this new right. The elections highlighted the fact that a considerable number of the Nisei in Japan had come of age while holding dual citizenship; young Nisei who were not of voting age were able to maintain their dual status. Many of the Nisei who lost their U.S. citizenship attempted to regain it by pressing for special legislation or filing lawsuits, a lengthy process filled with complex technicalities.

The first American report about Nisei desiring to return to the United States was filed by an NBC Tokyo correspondent on February 27, 1946. "The Japanese Americans who have been forced to remain in Japan since the outbreak of the war," he reported, "will be allowed to return to the United States upon verification in their records that they did not collaborate with the Japanese military during the war."[12] This report was based on the fact that some three hundred to five hundred Nisei residing in Japan had appealed to MacArthur's GHQ for permission to return to their homeland.

Within the State Department, proposals were being drafted to manage the repatriation of the Nisei, including a plan to establish a special hearing committee within the occupation to determine the eligibility of Nisei for reentry to the United States. The NBC broadcast noted that there was some question whether the investigation of the activities of the Nisei during the war was constitutional, but the U.S. government could hardly overlook the activities of Nisei living in enemy territory during the war.

On May 8, 1946, GHQ ordered the Japanese government to compile and submit to American authorities lists of all Japanese Americans residing in Japan during the war, those who had obtained Japanese citizenship, and those who had served in the Japanese military or government institutions. It was apparent that GHQ would decide the eligibility of each Nisei to reenter the United States on the basis of these lists.

Provisions were apparently made to exempt individuals from charges of rendering service to the enemy if it was proved that the actions were taken under duress. Thus the surviving Nisei women who intercepted shortwave radio broadcasts for the special intelligence group in Hiroshima took immediate steps to obtain written proof that they were forced to perform this work. As early as fall 1945 and into spring 1946, they made contact with Lieutenant Colonel Kakuzo Ōya and Major Kumao Imoto, two of the officers in charge of their group, to obtain the necessary documents.[13]

The startup of the eligibility hearings faced numerous delays in the latter half of 1946, but eventually the pace quickened and by early 1947 it took only three weeks from the filing of an application to the hearing. On August 15, 1946, five Nisei disembarked at San Francisco from an American vessel, the SS *Flying Scott*, signaling the first postwar entry into the United States of Nisei who had spent the war years in Japan. The five included three Nisei women, a Reverend Ueda from Santa Rosa Catholic Church in California, and a twenty-six-year-old man named Tomoya Kawakita.[14] Kawakita was the second son of Yasaburo Kawakita, who had made a fortune with the Kawakita Trading Company in the Mexican town of Calexico, near the California border. Tomoya had arrived in

Japan in 1939 to study at Meiji University in Tokyo and was unable to return to the United States when the war started.

Commercial traffic between the United States and Japan was resumed on February 23, 1947, by the President Line's SS *Marine Ada*, sailing from Yokohama. Five Nisei girls were listed among its passengers. It was thought that about nine hundred Nisei residing in Japan were waiting for transit. Fifty Nisei were on board the second ship to arrive, SS *General Gordon*, which docked in San Francisco on March 19, 1947. So far no hibakusha were among the passengers.

According to newspaper accounts, the first Japanese American hibakusha to reenter the United States was a young boy named Tohru Nishikawa. He was a passenger on the SS *General Gordon* when it arrived at San Francisco on May 16, 1947. *Rafu Shimpō*, the Los Angeles newspaper, reported on June 2, 1947: "Tohru Nishikawa, the eldest son of Kazushige Nishikawa, who owns a watch shop in the Japanese section of Los Angeles, returned here with a scar on his neck from a wound he received from flying debris during the atom bomb blast."[15] Tohru was interviewed by a reporter in Los Angeles for the article. "That morning I was in Kannon-machi," he said,

> a short distance from the epicenter. Members of volunteer groups from the suburbs, who were helping to evacuate buildings, were pulling carts, hurrying on their way to knock down houses in the city. Female students, wearing red hair bands, had also been mobilized and were walking to their work assignments in the city center, while students working the night shift made their way home looking exhausted. That was when I saw the flash, followed by a thundering blast that assaulted my eyes and ears. I ran frantically in the direction of Koi railroad station. This is the injury I received (Tohru pointed to his neck) from a broken window while I was running away. What I saw still terrifies me. Many were crushed to death underneath buildings, or overcome by burns, or had their lungs blown out by the blast. It was a living hell. . . . On the third day after the blast the fires subsided in the city, but forest fires [in the hills near the city] continued to burn, causing intense heat that killed many of those injured by the bomb. Over the next few months many people died of blood problems, burns, and gangrene.

This was the first eyewitness account of Hiroshima to be brought to the Japanese community in America.

On June 6, 1947, American newspapers reported under banner headlines that the FBI had arrested Tomoya Kawakita the previous day on suspicion of cruelty toward American soldiers at the Ōeyama POW camp during the war.[16] Kawakita, who had returned on the *Flying Scott*, had been conscripted after his graduation from Meiji University and assigned to the POW camp as a prison guard because of his knowledge of English. A clerk at a Sears Roebuck store in Los Angeles where Kawakita was shopping happened to be an ex-POW and identified him. The trial proceedings brought to light the way that Kawakita, born and raised in California as a Nisei, had mistreated American POWs. Kawakita's arrest revealed the complicated circumstances and the complex psychological situation of the Nisei living in Japan during the war, and it had a tremendous impact upon the Japanese community in the United States.

Kawakita was sentenced to death for treason. His sentence was commuted to life imprisonment by President Eisenhower, and after serving some years at Alcatraz, he was deported to Japan by executive order during the administration of President John F. Kennedy.[17] Those who were most disturbed by the implications of the *Kawakita* case were the associations of Nisei veterans who served in the U.S. forces during World War II. They issued a statement demanding a thorough investigation of the loyalty of the Nisei who wished to return to the United States from Japan. They also expressed their full support for the disclosure in U.S. newspapers one month prior to their departure from Japan of the photos, personal histories, and fingerprints of Nisei who had been granted permission to enter the United States.[18]

This reaction was considered entirely appropriate among those Nisei in America who wanted to be "120 percent American." On the other hand, Nisei living in Japan must have blamed fate for making them suspect in the eyes of their countrymen, and for the role the state played in trifling with the fate of individuals. Despite the public emergence of these volatile issues, however, the return of Nisei to the United States continued expeditiously even after Kawakita's arrest. It was rather common to see trains leaving Hiroshima station, which was only partially rebuilt at that time, packed with people on their way to Kobe to see off Nisei who were returning to America.

Fifty Nisei were among the passengers on board the SS *Marine Lynx* of the President Line, which anchored in San Francisco on June 6, 1947. Among them was Sadako Obata, who was pregnant when she experienced the atomic bomb in Hiroshima. She boarded the vessel after giving up her baby daughter, who was exposed in utero, for adoption. She landed alone on the shores of her homeland after an absence of seven years. The *Rafu Shimpō* noted her return to the United States with the following headline: "Miss Obata—With Deep Scars of Atomic Bomb Burn."[19]

Notes

1. Interview with Irene (Ishigame) Nakagawa, April 3, 1978, Los Angeles.

2. On the history of the Military Intelligence Service, see Tad Ichinokuchi, *John Aiso and the MIS: Japanese Soldiers in the Military Intelligence Service, World War II* (Los Angeles: MIS Club of Southern California, 1988).

3. Bradford Smith, *Americans from Japan* (Philadelphia and New York: Lippincott, 1948), p. 325.

4. Interview with Sadako (Obata) Shimazaki.

5. Interview with Judy Enseki.

6. *Rafu Shimpō*, February 4, 1946.

7. Shogo Muto, "Nisei no kotodomo" (On Niseis), *Rafu Shimpō*, January 29, 1947.

8. Interview with Irene (Ishigame) Nakagawa.

9. Toshio Sumita, "Kaette kita nisei shiminken hokisha" (Nisei returnees abandoning U.S. citizenship), *Rafu Shimpō*, July 29, 1946.

10. *Ibid.*

11. *Rafu Shimpō*, March 22, 1947.

12. NBC radio report reported in *Rafu Shimpō*, February 28, 1947.

13. Interview with Kakuzo Ōya.

14. *Rafu Shimpō*, August 1946.

15. *Rafu Shimpō*, June 2, 1947.

16. For example, see *Los Angeles Times*, June 6, 1947.

17. On the *Kawakita* case, see the U.S. Supreme Court decision, No. 570, June 22, 1952.

18. *Rafu Shimpō*, June 11, 1947.

19. *Rafu Shimpō*, June 9, 1947.

six

Strangers in Their Own Homeland

The number of Nisei leaving occupied Japan to return to the United States reached its peak in 1948. A history of the southern California Japanese published that year reported that the number of Nisei returnees had reached three thousand.[1] Since this figure was only for southern California, where there was the largest concentration of Japanese Americans on the mainland, we might estimate the total figure for the mainland United States at roughly five thousand. There is no way to estimate how many of these were hibakusha with scars, in their hearts or on their bodies, from the atomic bombings of Hiroshima and Nagasaki (although there would have been far fewer of the latter).

Nisei chose to seek a new life in the United States for reasons that were often complex. America was their homeland, but to those who had spent their formative years in Japan, it was nevertheless a foreign country. In the hearts of many, Japan was still their spiritual home. Still, many chose to return to the United States, a choice they clearly might not have made had the United States not won the war.

Kanji Kuramoto, a Hawaiian-born Nisei, escaped the direct impact of the atomic bomb, because he had been dispatched to a navy yard in the nearby city of Hikari under a student mobilization program during the final period of the war. After the bombing he entered Hiroshima, and for two weeks he searched for his father, who was assumed to have been bombed in the Ushida-machi district of Hiroshima. During those weeks, he thought he had seen actual scenes from hell. "Returning to the United States," he later testified before a California State Senate hearing in May 1974, "was an escape from the tragic experience and gave me great relief."² His younger brother, Tokuso, actually experienced the atomic bombing. When Kanji decided in June 1948 to return to the United States, relying on an uncle in San Francisco for help, he felt "a faint sense of hope, something like adolescence." Behind his hope for life in the United States, however, lay despair for Japan's situation and future.

Using the framework Robert Lifton developed to understand the psychology of Hiroshima A-bomb survivors in his book, *Death in Life*, one might assume that the Nisei hibakusha could no longer endure the guilt of living in Hiroshima, which had been devastated by a bomb dropped by their own country.³ However, I have never encountered a survivor who articulated his or her feelings in these terms. The reasons that prompted Nisei to return to the United States seem to have been more pragmatic. Kuramoto admits that he wanted to repatriate rather than to stay in "the totally devastated, hopeless situation of Japan."⁴ Akira Furuta also left for America, having decided, "It was no use to remain in Japan, where there was no hope." Thirty years later, he admitted that he "wouldn't have come if Japan had been then as it is today."⁵

Certainly educational opportunities were far more abundant in the United States, since it was hard even to buy a pencil and a notebook in postwar Japan. Patriotic loyalty to the United States likely played little role in Nisei deliberations, although the question of that loyalty would come into play when a Nisei boy reached the age of eighteen and became eligible for the draft. Imagine a young Nisei hibakusha, summoned to the U.S. Consulate in Kobe and asked to declare allegiance to the United States. You would hardly expect him to be filled with patriotic resolve to defend the country. But refusing to do so would mean relinquishing the possibility of returning to the United States.

Furuta, who had been injured by the atomic bomb, did not even think about the question of allegiance until he was drafted during the Korean War. "Why do I, who was almost killed by the American bomb, have to fight as an American soldier?" he wondered. However, he persuaded himself that fighting for the United States was the price he had to pay for choosing to live there.

Some Nisei returned to the United States after the war to deal with property disputes arising from the forced evacuation of Japanese families from the West Coast. Mitsuko Mizuno (not her real name), a Nisei woman who had gone to Japan a year before the war with her family, married a Japanese student who was not an American citizen. She suffered a serious injury to her leg from the atomic bomb, which she experienced with her children near Yokogawa Station in Hiroshima. In August 1947, Mitsuko returned to Los Angeles with a very young son and a nine-year-old daughter, Kayako, who had keloid scars on her arms. The land the family owned in California was to be confiscated if it was left unclaimed, and she was the only one able to reenter the United States to settle the issue. It was her original plan to return to Japan as soon as the matter was resolved, but the case was not completed until 1956. In the meantime, her other sons joined her in America, eventually followed by her husband once he succeeded in obtaining permanent resident status. Thereafter, the family has lived rather happily in the United States without thinking much about the A-bomb. The experience of the atomic bomb, her daughter commented, "was a big incident in my life, but I don't think it really affected me that much. It just happened, I feel. I don't think too much about whether the U.S. was wrong or right."[6]

Irene Ishigame also returned from Japan, after an absence of eight years, to recover her family's property in Fresno, California. Irene and her sister Jane, it will be recalled, were living in Japan during the war. Irene's family was sent to Jerome Relocation Center in Arkansas after the war broke out; her father was separated from them and sent to Tule Lake after he refused to take the loyalty oath. When the war was over, he stubbornly clung to the belief that Japan had won and returned there the year the war ended; the rest of the family joined him in Japan in 1946. Irene was young, but she was an adult U.S. citizen, and it therefore fell to her to recover the family property in California. It had been only two years since Irene had gathered up the ashes of colleagues who were bombed while intercepting radio messages in Hiroshima. The memory of the atomic bomb explosion was still vivid in her mind. While she was staying in Los Angeles awaiting settlement of her case, she married a Nisei and decided to stay on.[7]

Motoko Shoji (not her real name) was born in California, but she was sent to her grandparents in Japan when she was only a year old. In August 1945, by then a young married woman, she was exposed to the atomic bomb in Hiroshima along with her small baby, who died. This tragedy was followed by the notice of her husband's death in battle. Even though she was an American citizen, Motoko had never thought of returning to the United States, especially after the war and the atomic bomb. This changed when she remarried a Nisei who had renounced his American citizenship right after the war. In the face of the hardships they endured in postwar Japan, her new husband rather belatedly decided to return to California. The only way he was eligible to do this, however, was as the husband of Motoko, who had retained her American citizenship.[8]

Each of these individuals had his or her unique history, but all shared the common fate of being American hibakusha. Most of the Nisei who returned to their homeland after the war, whether or not they were hibakusha, shared a determination to retrieve the American identity they had lost during the years in Japan, to assimilate into the life of the land from which they had been separated. But was the America they returned to really a homeland? Did their country understand and embrace those who had been away for many years and who had suffered through the atomic cataclysm? Before this can be answered, we need to first look at another group of people who also "returned to America."

When some 120,000 Japanese nationals and American citizens of Japanese ancestry were evacuated from the West Coast and sent to relocation camps in the desert, their children were especially perplexed. They could not understand why they were stuck in a place populated entirely by Japanese. "Mommy, let's go back to America," was a frequently heard plea. For, although the Japanese community in America had certainly been cohesive, people of Japanese origin had always mixed with other ethnic groups, especially in the cities. The Japanese had formed ethnic neighborhood communities like Little Tokyo in Los Angeles, and the Japan Towns of San Francisco and Seattle, but those who lived in such communities were a small minority compared with Japanese who lived elsewhere. Cohesiveness among Japanese had been defined more by a spiritual and cultural identity than by physical proximity.

Ironically, for many Issei, life in the wartime camps was in some ways a long overdue "vacation," the first many of them had ever taken. Not discounting the fact that they were confined behind barbed wire, the internees did not have to worry about earning a living and there was plenty of free time. Japanese culture flourished in the camps, which were almost like Japanese oases in the middle of the American desert. And when, toward the end of the war, the government encouraged internees to leave the camps, rather few of them took the opportunity to do so. To be sure, this was a perverse and enormously expensive "vacation." In addition to the psychological damage done to the internees, material losses caused by the forced evacuation of the Japanese have been estimated at $400 million.[9] There is no way of estimating what could have been earned from their farms and businesses during those one thousand days of internment had they been allowed to continue as productive members of society.

As the war drew to a close, all of the camps were shut down, and the internees had to "return to America." They were advised not to return to the West Coast, where anti-Japanese sentiment remained strong. As a result, those leaving the camps dispersed to cities across the country where they had no deep previous ties: 3,000 went to Denver, 2,000 to Salt Lake City, 11,000 to Chicago, 16,000 to Detroit, 13,000 to Minneapolis, 600 to Cincinnati, and 3,000 to Cleveland. In March 1946, Tule Lake was the very last of the relocation centers to close. As of that date, some 57,000 returned to the West Coast, while roughly 52,000 moved to other areas of the country.[10] (Quite a few of those who moved from the camps to the interior of the country and the East Coast eventually returned to the West Coast, especially to California).

Psychic wounds remained in the hearts of those who had endured officially sanctioned ostracism and a concentration camp experience in their own country. "A bitter evacuation legacy shared by ex-inmates in varying degrees," says Michi Weglyn, "is a psychic damage which the Nisei describes as 'castration,' a deep consciousness of personal inferiority, and proclivity to noncommunication and inarticulateness, evidenced in a shying away from exposure which might subject them to further hurt."[11]

Many Nisei who experienced the camps during their youth avoided discussing it for years thereafter. Only in the late 1970s and 1980s did they begin to talk about it, often at the urging of their Sansei (third-generation) children, who were trying to learn more about what happened during the war in order to challenge the official version of recent American history. The Sansei were struggling to establish a collective identity as Japanese Americans, but also wanted to transcend the provincialism of the Japanese community and identify themselves as Asian Americans. Sansei created a "Japanese American Pilgrimage" and many visited the sites of Manzanar and Topaz, which their parents had refused for decades even to mention. Since they did not experience the camps, the Sansei were spared the psychic scars that their parents acquired as internees.

The late Edison Uno, a Nisei historian, once explained the unwillingness of the Nisei to speak openly about their camp experience in the following way: "We were like the victims of rape. We felt ashamed. We could not bear to speak of the assault, of the unspeakable crime."[12]

Nisei compensated for their sense of shame during the war not by speaking out, but by taking action. Many Nisei from the relocation centers volunteered to serve in the U.S. Army's 442nd Regiment, a combat team that distinguished itself in Europe under the slogan "Go for Broke." In the history of the U.S. Army, the Nisei unit received the highest number of decorations, and suffered the highest ratio of casualties. Nisei proved their loyalty to the nation with their own blood, and contributed significantly to turning around the prevailing negative image of the Japanese in America.

Today, Japanese Americans are called a "model minority." Many Nisei have been diligent and hard working, thanks to the strict discipline of their Issei parents. Their crime rate is extremely low, and the percentage of those who pursue higher education is extraordinarily high. The Sansei seem to have inherited many of the same values. The postwar policy of friendship between Japan and the United States and the amazing economic recovery of Japan have also contributed to the improved image of Japanese Americans. They are the American minority "who made it" and "who are accepted." However, they had to make extraordinary efforts after the war to gain such acceptance.

One hibakusha who returned to the United States shortly after the war says he was shocked to find how poor the Japanese American community was at that time. "The American dream" was not to be found there. The evacuation had robbed Japanese Americans of just about everything they owned, and the meager compensation they received from the government in 1948 did not cover even one tenth of their actual material losses. Japanese Americans internalized the psychic wounds they had suffered in the relocation camps, and even used this as an inner engine to drive them to success in postwar America. Despite formidable obstacles, they rebuilt their economic base on a stronger foundation than had existed before the war. The rapid economic growth of postwar America certainly helped, but their achievement also reflects decades of hard work.

Those who came back from Japan as hibakusha were also left with psychic wounds. In their case, the crisis of identity was compounded by having witnessed the "hell" of the atomic bomb. Although they were often members to the same family, those who lived through relocation and those who endured the atomic bomb experienced two different wars. Even when they were reunited, it was very difficult for them to understand each other. The Nisei hibakusha became a minority within the Japanese American minority, and they often felt isolated and alienated.

Most of the younger Nisei who spent their formative years in Japan had lost their proficiency in English and faced a severe language problem after they returned to America. For many, English had become a foreign language. This created enormous problems among the returnees, and many of them fell into a cultural chasm between the two countries. Even decades later, these individuals suffered from a mental torment that Dr. Thomas Noguchi, then Los Angeles County Medical Examiner-Coroner, defined as "cross-cultural stress."[13] Such alienation was felt especially keenly by women, who generally had fewer opportunities for assimilation into American culture. A woman could remain forever alienated through confinement within the cocoon of the Japanese community. It was not unusual for a Kibei Nisei woman to live out her whole life in the United States speaking hardly a word of English.

A large majority of the Nisei hibakusha carried this severe language handicap. They thus found it exceedingly difficult to discuss their atomic bomb experience with other Americans, who were predisposed to believe that the use of the atomic bomb had been justified on military grounds (or who harbored the sentiment that the bomb was a just reprisal for the Japanese attack on Pearl Harbor). Hibakusha in the United States have less difficulty communicating their experiences in Japanese, but even then often find it terrifying to assert themselves, or even to speak about the inhumanity of the bomb. Most American hibakusha became introverted.

The story of Kaz Suyeishi, a Pasadena-born Kibei who went from Japan to Hawaii to study dress-making before returning to mainland America, helps illuminate the situation facing the hibakusha upon their return to the United States. Doctors in Japan had at one time given up on Kaz because the effects of radiation exposure made her physical condition so precarious. By the time she arrived in Hawaii, however, she seemed much improved. "Then, about a year after I came to Hawaii," Kaz recalled,

> large colored spots began to appear on my arms, and I had a high fever and suffered weight loss. I thought it might have something to do with my being away from home. My doctors told me that I was homesick and should go back to Japan. After a month of rest on another island, however, I recovered.
>
> After returning to Honolulu, a white man in his fifties pointed his finger at me and said, "You bombed Pearl Harbor and killed our boys." That really reminded me that I was a victim of the atomic bombing and I wanted to tell him that, but my English wasn't good enough. I struggled to communicate to him with all sorts of gestures, but after that experience I became depressed. I developed a rash all over my body. Dermatologists and internists all kept saying it was because I was homesick. My parents wrote and told me to come home but I had come all the way to Hawaii at great effort. I was determined to see my birthplace and even go to college in America.
>
> In 1952, I did enter college in Pasadena, where I had been born. I was told that my English was the worst in the history of that college. At that time I was physically OK, but within six months I was once more back in the same condition. As a part of my speech class, I had to go to different places and in my broken English introduce myself to many Americans. When I said I came from Hiroshima they all asked me about the atomic bomb. At first, I was a little hesitant, but I told them honestly about my experiences. During the daytime, I wasn't consciously upset, but every night I had nightmares. I panicked. At first, my hands and feet were numb; then, I had a pressing sensation around my heart and difficulty in breathing. These were the same symptoms I'd had in Hiroshima. Gradually, I became used to it, but I became depressed again. Three days out of five, I had to rest. This lasted for about a month, and finally, in 1955, I returned to Japan.[14]

There was a brief period of time—during the two or three years immediately after Hiroshima and before the Cold War with the Soviet Union intensified—when many conscientious Americans regretted the atomic bombing. During that period, the United States had a monopoly on atomic weapons, and Americans were still capable of exhibiting large-mindedness as victors in the war. It was also a time when many scientists, including Oppenheimer, who had participated in the development of the atomic bomb returned to their universities and labs and lent their voices to warning about the perils of nuclear war.

The phenomenal, emotional response to *Hiroshima*, John Hersey's book based on the testimony of six survivors in Hiroshima, was indicative of the American openness of that period.[15] Published in the *New Yorker* magazine August 31, 1946, a year after the bombing, it occupied the entire issue. On the day of publication, the unprecedented issue sold 300,000 copies in New York city alone. Albert Einstein, who had advocated production of the atomic bomb, bought one thousand copies to distribute among his friends.

Hibakusha who were trying to overcome the misery and horror of the atomic bombing contributed to this critical reappraisal of the meaning of Hiroshima. For example, the Rev. Kiyoshi Tanimoto of Nagarekawa Church, one of the survivors featured in Hersey's essay, was invited to the United States in September 1948 by the Mission Board of the United Methodist Church. For fifteen months, Tanimoto traveled extensively throughout the United States sharing his testimony about the horror of the atomic bomb and asking for funds to build a peace center in Hiroshima. In the journal of his peace tour, he reports the experience of being told by an American intellectual, "Forgive us for the atomic bomb. It was the most outrageous crime that was ever committed in the history of mankind."[16]

It was also around this time that Rev. Allan Hunter of the Mount Hollywood Congregational Church in Los Angeles made a cross from the burnt wood of a camphor tree left standing on the grounds of the Nagarekawa Church in Hiroshima, which he placed on the altar of his church. At the dedication service on November 27, 1948, Eiko Kakida, who had brought part of the tree from Hiroshima, led a procession with the cross, and Tanimoto delivered "A Message from Hiroshima." Kakida, a Nisei born in Bakersfield, California, moved to Hiroshima before the war. Her father was killed by the atomic bomb, and she returned to California after recovering from a severe mental breakdown as a result of her own radiation exposure. She was living in Los Angeles as Eiko Ohnishi and recalled "the joy of having been able to share with American people the true misery of the atomic bombing."[17]

As the Cold War intensified, however, American public opinion began to shift. The accounts of American survivors of Hiroshima were no longer heard. Hibakusha who returned to their homeland had to begin their new American lives in this changing climate.

In September 1949, the Soviet Union successfully conducted its first atomic-bomb test. Suddenly, the American monopoly on atomic weapons had been broken. In January 1950, President Truman ordered the Atomic Energy Commission to launch the production of hydrogen bombs, and in June the Korean War began. America started to lose its sense of self-confidence. Communist gains throughout the world had begun to shake American composure.

When Rev. Tanimoto visited the United States again in September 1950, he reported that the editor of *The Christian Century* summarized the drastic change of climate in America in the following way: "Sympathy toward Hiroshima and conscientious objection to the use of nuclear arms have now disappeared. America is now considering where we might use our next atomic bomb."[18] Once again, in the course of his travels, Tanimoto shared his atomic-bomb experience. Now, he found, his audience wanted practical information on how to defend themselves from an atomic bomb attack, and how to survive a bombing.

The anxiety of the American people was only intensified by the civil defense plan issued by the federal government as a national policy. Its purpose was to urge people to build fallout shelters in order to minimize damage in the event of nuclear attack. Subway tunnels and basements of large buildings were designated fallout shelters, and ominous signs with three black triangles against a yellow background were displayed on every building so designated. Evacuation drills were announced by sirens and people built shelters in their back yards. The shelter boom became a symbol of the frantic fifties. The first shock wave of that decade came in the period between the outbreak of the Korean War and the Russian hydrogen bomb test in 1953, and a number of succeeding waves followed in the United States as the international situation grew worse.

In the 1950s, the only lesson America seemed to have learned from the experience of Hiroshima and Nagasaki was that if a nuclear attack could be detected in advance, and if adequate fallout shelters could be prepared, a majority of the people would survive. This was the basic assumption of the civil defense program. Brigadier General Crawford F. Sams, Chief of the Public Health and Welfare Section at MacArthur's GHQ, was one of the most knowledgeable people on the subject of the destructive power of the atomic bomb, for it was he who had surveyed the effects of the atomic bombing in Hiroshima and was responsible for the definition of "death by atomic bombing." After retirement, he himself built a shelter with a hand-operated ventilation system in the backyard of his northern California home.

In the early 1950s, the vivid memory of the mushroom cloud was still fresh in the heart and mind of Jack Dairiki, who as a student in Hiroshima at the time of the bomb had been mobilized to work in war-related industry. He later returned to Sacramento, his hometown, and in 1951 was studying architecture at a local junior college when he recreated his memory of the atomic bomb for an art class. The drawing—a giant mushroom cloud in the distance behind his factory—intrigued his art teacher, and a reporter from the local newspaper took an interest in it. A story, headlined "This is Hiroshima—A Young Sacramento Nisei Speaks of the Horror of the A-Bomb," appeared in the Sunday edition of the paper, along with Jack's drawing.

Given the tenor of the times, American officials easily incorporated this eyewitness account of the bomb into the campaign for civil defense. The following year, the California Department of Civil Defense produced a radio program that featured an interview with Jack Dairiki.

Moderator: Countless Californians are inclined to believe that they are doomed to destruction should an atomic bomb be exploded over the city in which they live. Major General Walter M. Robertson, state director of civil defense, has repeatedly stated that this is not so. He has often given assurance that many will survive *if* they simply observe the fundamental rule of survival: "Take cover and stay put." This broadcast will offer living proof that General Robertson is right. Here with me now is Jack Dairiki, a twenty-one year-old student of Los Angeles City College, who was at Hiroshima that fateful August 6, 1945. Tell us, Jack, what is an atomic bomb explosion like?

Dairiki: The people of Hiroshima have a popular expression to describe it—"Pika-Don."

Moderator: And just what does that mean?

Dairiki: It's a somewhat childish expression. But "pika" refers to a strong, severe flash. And "don" means what you might think—a great, loud noise.

Moderator: Where were you when the bomb fell, Jack?

Dairiki: It was just by luck that I wasn't right in the middle of the city. That morning the train from Osaka, which I took each day to get from my grandfather's farm to the factory where I worked, was thirty minutes late. By the time I got to the factory there had already been an air raid warning in Hiroshima City. But no planes had come over and the all clear signal had been sounded. So I changed into my work clothes and was already outside the factory ready to go into the city when I heard planes coming. I looked up and could see three B-29s going directly over Hiroshima City. Then suddenly there came a bright flash of light. It was just like looking directly into another sun—it was so bright and yellow color. I started to run for the open doors of the factory. Maybe I took three, maybe four steps—I don't remember—but suddenly there came the noise of a big bang. When I heard that sound I threw myself flat on the ground and covered my eyes and ears with both hands.

Moderator: That's what you had been trained to do—cover your eyes and ears with your hands while you lay flat on the ground?

Dairiki: Yes. In that way we had protection from the bomb explosion blast, and it wouldn't break our eardrums or make our eyes pop out. Everybody had been taught to lie down flat right away and protect their eyes and ears. I lay on the ground about ten, maybe fifteen seconds, while the terrible wind from the bomb explosion broke all the windows in the factory. Everything was flying through the air. Broken glass, pieces of metal, boards—everything was falling all around me as I lay on the ground.

Moderator: What did you see when you got up from the ground, Jack?

Dairiki: [At first] I couldn't see anything. The air was so full of dust nobody could see. But I knew where the cave shelter was located on higher ground. So I ran to that shelter and others came running there too. When I got up this slope to the entrance to the shelter, I was able to see over the dust as I looked toward the city.

Moderator: Won't you tell us, in your own words, just what it looked like?

Dairiki: The first thing I saw was this tall column of smoke. It seemed to have all of the colors of a fire. And on top of this column of smoke was spreading out this big cloud with the same colors of fire all through the cloud. I was scared—everybody was scared, I guess—so we ran into the cave shelter.

Moderator: What about civil defense in Japan during the war, Jack? Did they have any?

Dairiki: Oh, yes. Very good. After the first air raid attack on Tokyo they began training the people in civil defense.

Moderator: That was the raid led by General Doolittle. What sort of training did the people have, Jack?

Dairiki: Everybody, even the children in schools, [was] taught how to take care of themselves in case they were hurt by bombs. They learned first aid. And everybody had to carry a first-aid kit with them everywhere they went. Another thing—all the people had to wear something like helmets to protect their heads. Many had regular helmets. The others, like in schools, had to make their own with cloth and cotton for protection. This was so people wouldn't get hurt by the pieces [of] antiaircraft shells falling on them.

Moderator: What about firefighting? Did the people have training in this too?

Dairiki: Oh, yes. Everybody, even the women, were trained to put out fires in their homes. Very few homes had anything like fire extinguishers. But they all had buckets filled with water and some filled with sand. And small rugs were kept rolled up and handy so they could be soaked with water and used to smother small fires quickly. Families practiced firefighting with regular drills and neighbors helped too.

Moderator: Do you think civil defense is important, Jack?

Dairiki: Oh, very important. It was the training I had in how to protect myself that kept me from getting badly hurt—maybe even killed—when the atomic bomb went [off] over Hiroshima City. If I hadn't dropped flat on the ground I could have been killed by the force of the blast. Knowing first aid saved many, many lives of the people of Japan. Red Cross and hospitals are all right in a little disaster, but when something big like the atomic bomb comes, it's up to everybody to take care of themselves.

Moderator: And the fire training and rescue training, did that help also when the atom bomb was set off over Hiroshima, Jack?

Dairiki: Many homes were saved by the people who lived in them because they knew how to put out small fires. It was very important in some parts of the city which would have been completely burned up if the people hadn't known what to do. And they saved many lives of people buried in wreckage of homes which had been knocked down by the force of the blast.

Moderator: Thank you very much, Jack Dairiki, for this first-hand account of how you lived through the experience of an atomic bombing at Hiroshima. The facts you have given us about the value of learning self-preservation and of taking part in mutual aid should help the people of California in preparing to meet atomic attack, if it should come to any of our cities.[19]

At the time of the Hiroshima bombing, Jack Dairiki was three and a half miles from the center of the blast. Within roughly a one-mile radius of the hypocenter of the atomic bomb, Japan's civil defense precautions were utterly without effect. Dairiki's comments in the interview may have reflected his limited experience of the impact of the bomb. They certainly reflected a tendency, often found among the Nisei, to accommodate to the expectations of those in authority and, in particular, to avoid the controversial aspects of the Hiroshima bombing.

Notes

1. Dōjun Ochi, ed., *Minami-Kashu nihonjinshi* (History of the Japanese people in Southern California) (Los Angeles: Nanka Nikkeijin Shōgyō Kaigisho, 1957), p. 743.

2. Record of Hearings at the California State Senate Subcommittee on Medicine, Education and Welfare, "Health Problems of Atomic Bomb Survivors," May 4, 1974. (Hereafter cited as State Senate Hearings).

3. Robert Jay Lifton, *Death in Life: The Survivors of Hiroshima* (New York: Random House, 1967).

4. Interview with Kanji Kuramoto, June 12, 1976, San Francisco.

5. Interview with Akira Furuta (pseudonym).

6. Interview with Mitsuko Mizuno (pseudonym) and her family, June 19, 1976, Los Angeles.

7. Interview with Irene (Ishigame) Nakagawa.

8. Interview with Motoko Shoji (pseudonym), June 5, 1976, San Francisco.

9. Committee to Investigate War Relocation and Detention, ed., *Individual Justice Denied* (1982).

10. Honkō Matsumoto, *Fukkōsenjō ni odoru kikan dōbō* (Fellow Returnees Active in Rebuilding the Life) (Los Angeles: Rafu Shoten, 1949), p. 128.

11. Michi Weglyn, *Years of Infamy* (Seattle: University of Washington Press, 1996), p. 273.

12. Maisie and Richard Conrat, *Executive Order 9066: The Internment of 110,000 Japanese Americans* (San Francisco: California Historical Society, 1972), p. 10.

13. Interview with Thomas Noguchi, June 17, 1976.

14. Interview with Kaz Suyeishi.

15. John Hersey, *Hiroshima* (New York: Knopf, 1946).

16. Kiyoshi Tanimoto, *Hiroshima gembaku to amerikajin: Aru bokushi no heiwa angya* (The Hiroshima Bomb and the American People: The Peace Pilgrimage of a Japanese Clergyman) (Tokyo: Nihon Hōsō Shuppan Kyōkai, 1976), p. 109.

17. Telephone interview with Eiko (Kakida) Ohnishi, April 29, 1978, Los Angeles.

18. Tanimoto, p. 139. Also his interview, May 21, 1978, Hiroshima.

19. California Civil Defense, Program No. 43, February 11, 1952, typed transcript.

seven

Pieces of the Jigsaw Puzzle

An estimated one thousand hibakusha were living in the United States at the end of the 1970s.[1] They came from many different backgrounds, and each trod his or her own road to begin a new life in America. Each was unique, like the individual pieces of a jigsaw puzzle, scattered across the country. Only the A-bomb experience united them.

About five hundred A-bomb survivors lived in California, centered in Sacramento, San Francisco, San Jose, and the Los Angeles area. Of these, around three hundred had been identified by name and address by the early eighties. An additional three hundred hibakusha probably resided in Hawaii.

I have suggested that my efforts to document the story of the American hibakusha can be compared to putting together a jigsaw puzzle, but there can be no hope of ever completing the picture. All that can be done is to trace the personal histories of individual survivors in as much detail as possible, and then put them together to produce the suggestion of an image. This image is complicated by the fact that many Nisei hibakusha returned only belatedly to the United States, while many hibakusha who eventually came to reside in America were born in Japan and only came to the United States after the war as marriage partners.

The number of Nisei returning to the United States from Japan peaked in 1948, and the number returning as hibakusha probably peaked about the same time. There was a considerable decrease in the number of returnees after 1952, the year the San Francisco Peace Treaty came into effect and the occupation ended. Even thereafter, however, Nisei whose departures were delayed for various reasons continued to return, and their numbers continued to include hibakusha who had the A-bomb deeply stamped on their minds and bodies.

∽ ∽ ∽

Kaname Shimoda was born in Sacramento in 1931. He was taken by his parents to Japan at the age of five. Later, as a student at the First Municipal Technical School in Hiroshima, he was exposed to the A-bomb and burned on the right half of his body. For three years he was hospitalized for treatment of aftereffects of radiation exposure. Still, Shimoda held a strong desire "to see with my own eyes the country where I was born." He finally made it back to his home country when he was twenty-seven by enlisting in the U.S. Army.[2]

At the time, Shimoda was running a business in Hiroshima, but it had become burdensome and he felt he wanted to "leave behind my life in Japan" and start afresh. All of his family had returned to Japan before the war, however, and he knew nobody in the United States who might help him. Since it was difficult to finance the trans-Pacific trip, it seemed like a "splendid idea" to volunteer for military service, which would enable him to discharge his duty as a U.S. citizen and, above all, to return to the United States at Uncle Sam's expense.

Shimoda completed all the procedures necessary to enter the Army at Kure, near Hiroshima, and was sent to Fort Ord in California in August 1957. He recalls that his Japanese friends railed at him at a farewell party. They called him a fool for going to the barbaric country that had harmed him. His older sister, herself an A-bomb survivor, could not understand her brother's decision, and it made her resentful that he was going to the country that, as she put it to him, "almost killed you." His sister had been born in Japan and had married immediately after the war. She lost her first child, a boy born four or five years later, from an illness that she believed was caused by her radiation exposure.

Shimoda's desire to see his birthplace was strong enough to overcome all of these objections. After he arrived in the States, he went three times to Sacramento to see where his former home had been. There, his father had been a successful grape grower. "I found a big tree and the shack where I was born," he later said. "I remember telling myself that my father had struggled there and that I would work even harder."

Shimoda made several visits to Japan after he was discharged from military service, but he settled in Pasadena, where he ran a successful business in gardening equipment and enjoyed a large circle of acquaintances in the Japanese community. "I am loyal to this country, he commented, "so I attend veterans' meetings; I owe my business to this country." But, he added, "I like Japan and have not decided to settle here permanently."

Shimoda's ambivalence reflected a complex state of mind. "Each time I go to Japan," he continued,

> I pay homage to the graves of the A-bomb victims. I feel apologetic toward friends of ours who perished crying "Down with America and England!" I feel embarrassed now for having joined the American military. I can explain myself to the living, but ...
>
> How do I feel about the A-bomb? Maybe it could not have been helped, but I think it should not have been dropped on noncombatants. I wonder why they did not drop the bomb on some actual battleground, instead of Hiroshima. I am sure that someday American history will be revised to admit that the dropping of the A-bomb was a mistake. After all, the forced evacuation of the Japanese is now acknowledged to have been unconstitutional.

Kiyoko Oda was another hibakusha whose return to the United States was delayed.[3] A resident of San Jose, she was born in the northern California town of Loomis and was sent to Japan in 1921 at the age of four to be brought up there. She stayed on in Japan and got married before the war. Oda came back to the States in 1955, her return delayed because she had lost her citizenship after voting in a Japanese election. As a result of postwar occupation reforms, Japanese women were given suffrage for the first time, and Oda, like many women, felt compelled to go to the polls "for fear of General MacArthur's scolding" or because "it was General MacArthur's command." Since she had been living as an ordinary Japanese, it is not surprising that she didn't think much of it when the neighborhood association delivered an election notice to her.

Legally, Oda had possessed "dual citizenship" from the time of her birth, but she was not particularly conscious of having been born in America. During the war she had never felt that Japan was at war with her home country. At the time of the A-bomb explosion, she was at home in Minami-machi, a little over a mile from the center of the explosion, with her two young daughters. She did not know that she had been A-bombed, of course; she thought only that an "incendiary bomb had exploded, and we have to extinguish the fires." Since they were inside the house, they were not exposed to heat rays, but Oda's children were covered with blood from pieces of broken glass, which pierced her face and arms as well. The kitchen windows were blown out and a sewing machine was knocked over. It was a miracle, she recalled, that they were not killed. Soon after, Oda walked about the city in search of a friend. All of her family, with the exception of her husband, who was away for military service, suffered from the aftereffects of the experience.

Oda had four brothers who had returned to the United States before the war after finishing middle school in Hiroshima. Her older sister, who remained with her in Japan, died without seeing America again. As many successful immigrants from Hiroshima had done, Oda's parents had built a house in their hometown in Japan, where they returned to live in retirement. After the outbreak of the war, all four brothers were taken to the camp at Tule Lake. This development delivered a deep shock to their father, resulting in his death. The brothers were members of what was called the "Banzai group," loyal to Japan, and they remained at Tule Lake until the end of the war.

With her brothers and numerous relatives living in the United States, it was not surprising that Oda decided to return to her native country. The procedures for her return, including the reinstatement of her citizenship, were started in 1947 and took nearly eight years. When her citizenship was restored, Oda was told that she had to enter the United States immediately, before Christmas 1955, or the order would expire. Thus, at the age of thirty-seven, Oda had to leave Japan very suddenly, and she found herself "afraid of going to America." Her resolution was strengthened when she realized that having citizenship meant she could take her mother to the United States to see her sons. What began as a temporary visit, as often happens, ended in the resettlement of the entire family.

Oda continued to suffer from A-bomb-related illnesses; both her white and red blood corpuscles were less than half the normal count. While she was in Japan, everyone told her that she was going to America to die because she tended to be confined to her bed more than she was out of it. But she left Japan with the rather bold hope that "this illness given by America will be cured by America." After she got to the United States, Oda helped one of her brothers in his family business, growing fruit on a hillside. "In the beginning" she said, "I had to work, which in fact spurred me on, and my life turned around 180 degrees. It was a life," she continued, "only a little better than that led by the Issei."

Her daughters, exposed to the A-bomb at an early age, grew up. Oda became a grandmother and worked as a domestic, while her husband grew flowers. Life stabilized, but the fear of illness related to A-bomb exposure remained. Oda's neighbors, she reported, were kind and sympathetic toward her A-bomb experience, which was a great comfort to her.

Hibakusha whose return to the United States was delayed lived through some difficult times, but some, like Shimoda and Oda, managed to put down secure roots. Given their prolonged stay in Japan, however, they tended to long for Japan. Others, who were born in Japan and later in life chose to come to the United States, longed even more intensely for the old country.

After 1952, normal travel between the United States and Japan was resumed. Among those coming to the United States during this time were considerable numbers of new immigrants (as distinct from returning Nisei), and some among them were hibakusha from Hiroshima or Nagasaki. The vast majority of survivors coming to the United States were female hibakusha, married to American men.

During the early postwar period, many hibakusha had married soldiers and civilians attached to the U.S. military. They were called "war brides" during the occupation, but it might be more appropriate to call them "international brides," since many married after the restoration of Japan's independence. Most of the soldiers who married A-bomb survivors were white, although not a few Nisei soldiers stationed in occupied Japan married hibakusha.

A larger number of hibakusha marriages were based on regional ties to Hiroshima. Despite the fact that the marriages were between Japanese and Americans, they were "international" in only a formal sense. Traditional parents often hoped to find their son a bride of their own choosing from their hometown. This desire was especially strong among Issei who longed for Japan, but they were unable to realize these wishes during the Pacific war.

The bias against interracial marriage remained strong within the Japanese community, in part a consequence of the prewar and wartime experiences of racial prejudice. In addition, there was for many parents a preference for brides from Japan, who were considered more pliant than Nisei women.

Los Angeles resident Tomoe Okai's story is rather typical:[4] "I was born in Hiroshima city," she related, "and experienced the A-bomb along with my former husband, who died the following June from aftereffects of the bomb. Later, a friend introduced me to a Nisei from Los Angeles, and I came to this country to marry him. He is my present husband. It was also his second marriage. He told me that he had been hoping the second time around to marry a woman from Hiroshima, because, he said, they are reputed to be patient." This preference strongly marks the male Kibei. Brought up in a culture different from that of the United States, and having developed Japanese sensibilities, they quite naturally wanted to marry women who had also been brought up in Japan.

The A-bomb experience was nonetheless a hurdle to overcome when it came to marriage. It was rumored that some women tried to conceal from their husbands the fact that they were exposed to the A-bomb, but almost no such cases have come to the surface. Since Japanese communities in America are still closely knit, and Japanese carefully investigate family histories prior to marriage, the A-bomb experience would be difficult to conceal. It is likely there were very few marriages with such secrets, with the possible exception of marriages between hibakusha women and men who were not Japanese or Japanese American.

If a woman's A-bomb experience was not an impediment to marriage, it was possibly because of her prospective husband's love for her. But it is also likely that there was an optimistic assumption that it was possible to ignore, or at least to suppress into subconsciousness, the A-bomb experience and its future consequences. The A-bomb was, without question, a horrific experience, but it was possible for many to convince themselves that it was behind them. Newlyweds especially would embrace the notion that they could overcome, by their joint efforts, whatever difficulties might arise in the future.

This was a psychologically complex matter indeed. Hibakusha often complained that U.S. doctors had no understanding of how the A-bomb can affect the human body, and that they were not even interested in learning. These doctors, in the eyes of many hibakusha, simply did not recognize the existence of A-bomb diseases. But if a doctor declared a survivor free of bomb-related health problems, the couple frequently chose to embrace a favorable report.

How did hibakusha women who came to the United States as marriage partners feel about their new country? In general, they expressed few ill feelings. Yoko Ueshima (not her real name), married to a Nisei farmer in central California, experienced the A-bomb when she was in the fourth grade of a girls' high school.[5] The left half of her body was burned. The corpse of her younger brother, then in the first year of middle school, was never found. Nonetheless, she declared that she had "no resentment towards America." Strong-willed, she had expected to stay single all of her life until she had an offer of marriage from a Nisei. She was inclined to accept "more because I was attracted to America than to marriage." Her father objected to her going to the "country that killed your brother," but she said her fascination with America prevailed. As an instructor of the tea ceremony and the koto, a musical instrument, she hoped to teach in her own individual style in the United States, free from her authoritative Japanese mentors.

Tamae Taniguchi (not her real name) was suffering from the aftereffects of the A-bomb when she came to the United States to marry a Sansei from northern California.[6] In her own words, "By going to the States I wanted to say good-bye to my past" and to all the troubles at home that arose from the family business. Compared to Japan, America was the "country of freedom." She said that she was not particularly conscious of the fact that America dropped the A-bomb. "It does not help to think about that. Maybe I'm just happy-go-lucky."

Kaz Suyeishi of Los Angeles also came back to the United States to marry, because she thought American society was freer.[7] She understood that life might be hard in the United States, but, as she put it, "I can keep my privacy and enjoy life." She asked herself, she said, which was better: "To make my home in Japan and worry a lot, or live in America, perhaps in poverty, but free and private?" She eventually chose the latter. Suyeishi left her marriage entirely in her parents' hands and married a Nisei of her father's choice. It is noteworthy that she did not intend to bury her ashes in the United States, as she put it. "My husband is a Nisei, but we wouldn't mind living in Japan if we could make a living there; if things get bad here, we can always go to Japan."

The hearts and minds of A-bomb survivors who came to the United States to marry swayed back and forth between Japan and their new home. Many got married and remained as permanent residents without ever acquiring U.S. citizenship, whether or not they were A-bomb survivors. They may have left a part of themselves in Japan, but there was little chance of going back, especially since their children grew up American.

This ambivalence was, of course, not limited to those who married ethnic Japanese. Keiko Lincoln (not her real name), who married a white engineer, was a two-month-old fetus when her mother suffered exposure to radiation from the Hiroshima bomb.[8] She was considered to be at high risk of long-term effects, and the ABCC in Hiroshima examined her extensively, administered IQ tests, and took X-rays of her entire body. "After all these years, I have found out why they did all this," Lincoln said, adding only, "I am not very happy about it." Although she suffered from a decrease of white blood cells and from anemia, she insisted, "I don't feel that I experienced the A-bomb. I do not have a basis for feeling hatred."

Lincoln came to California to study after graduating from high school in Hiroshima. She had no special admiration of America, she said, but had wanted to study in the United States ever since junior high school. She married her husband after finishing junior college in California. By her own admission, Lincoln was not entirely comfortable with the American way of doing things, but her criticism of contemporary Japan was more severe than that uttered by any other hibakusha I met. Torn between Japan and the United States, she gave the impression that she was unable to resolve her own ambivalence.

Chizuko Favatella experienced the A-bomb when she was an advanced-course student in a girls' high school in Nagasaki.[9] She met and married her husband, an American high school teacher, when he was stationed with the medical corps at the Sasebo Navy base, where she worked as a clerk-typist. Favatella describes their love as "passionate enough to make me come along with him to the United States," and, indeed, she was happily married with three children. She felt that the A-bomb "couldn't be helped, since it was a war." Favatella commented, "I, personally, have no hatred toward America." By her own account, this attitude had something to do with the fact that she was raised in the cosmopolitan city of Nagasaki. An aunt had married a Frenchman before the war. This may have prepared her to go abroad; love might have conquered the rest.

Why then did Favatella, who was so well-suited to living in America, suffer violent bouts of anxiety? One certainly cannot attribute every such psychological (or physical) problem to the A-bomb experience, but perhaps all hibakusha harbor, deep in their hearts, a shadow of that cataclysmic experience.

"I gave birth to three children," she related,

> one after another, two girls, followed by a boy. I was hoping to have another boy when my health deteriorated. I had three miscarriages, and then I began having dizzy spells. I became terribly anxious. My marriage was close to ideal, and I was happy with my husband. We weren't bad off economically, and I had nothing to worry about, but I was seized by anxiety and by feelings of apathy. I was seeing a doctor all the time. I told my doctor about my A-bomb experience, but American doctors don't take that seriously. After seeing the results of my tests, he told me there was no connection to the A-bomb.

At the time I interviewed her, Favatella was feeling much better. She had, she said, "come out of a long tunnel by whipping myself. Even today," she continued, "my ears ring and I get terribly dizzy, but I sit still and tell myself that I will feel better. I try to think that it has nothing to do with the A-bomb. It's better to forget."

If only they could forget, everyone would prefer to do so. But Favatella's two daughters also tired easily, tended to feel dizzy, and had frequent headaches. "I don't want to blame the A-bomb for all this," Favatella said, her suppressed anger surfacing, "but if my daughters have inherited my A-bomb experience, I will surely feel bitter about it."[10]

Another example is Kuniko Jenkins, who experienced the A-bomb when she was working as a nurse at the First Army Hospital in the Motomachi district of Hiroshima.[11] It took her three years to recover from the nearly fatal effects of the bomb. She subsequently worked at the U.S. Army hospital in Tokyo, where she met and married her husband, a medical technician. After her arrival in the United States, she was seriously ill several times and pulled through only with her husband's constant love and care. Her face bore deep scars from the A-bomb. Jenkins's white blood cell count was far below normal, and she needed to carry an oxygen tank with her wherever she went. She was exceptionally lucky to be married to a man who was in the medical appliance business and through whom she could receive veteran's medical benefits. She showed her appreciation by becoming an American citizen.

"We A-bomb survivors have gone sour; we have seen hell on earth," Jenkins commented. Despite her poor health, she was active as a leader of a group of hibakusha in the San Francisco area. Her motivation seemed to stem less from anger than from the wish to help those who, like herself, suffered from the A-bomb.

There are certainly more pieces to the jigsaw puzzle than I have presented here. For example, of the twenty-five "Hiroshima Maidens" who were brought to New York in 1955 to obtain plastic surgery on their scarred faces and hands, three chose to remain living as permanent residents in the United States, while one lived in Canada. One was in California, married to a Nisei; another in Maryland, married to a white American; yet another, Shigeko Sasamori, was living in the Los Angeles area, divorced after having been married to a Japanese.[12] Sasamori, in sharp contrast to many of the hibakusha in America, was wholly engaged in serving as a living witness to the misery caused by the bombs, and determined to appeal at every opportunity for the abolition of nuclear weapons. The importance of these women to an understanding of the situation of hibakusha in the United States is beyond question.

Others—probably more than half of the estimated one thousand hibakusha in the country—chose not to make themselves known as A-bomb survivors. These "silent hibakusha" are also an important part of the puzzle.

A Korean hibakusha living in Nagasaki once said, "Those who felt the greatest bitterness are all dead."[13] Some of the stories we need to dig out and record as well as possible are the stories of the dead, such as the one that follows.

Notes

1. *American Atomic Bomb Survivors: A Plea for Medical Assistance* (Alameda, California: National Committee for Atomic Bomb Survivors in the U.S., 1979).
2. Interview with Kaname Shimoda.
3. Interview with Kiyoko Oda.
4. Interview with Tomoe Okai, June 17, 1976, Los Angeles.
5. Interview with Yoko Ueshima (pseudonym), September 4, 1977, California.
6. Interview with Tamae Taniguchi (pseudonym), June 18, 1976, California.
7. Interview with Kaz Suyeishi.
8. Interview with Keiko Lincoln (pseudonym), June 5, 1976, San Francisco.

Pieces of the Jigsaw Puzzle

9. Interview with Chizuko Favatella, June 18, 1976, Hollywood.

10. To date, research on the effects of radiation exposure has not identified any genetic complications that would account for symptoms in the children of hibakusha. There was much uncertainty attached to this issue in the decades after the bombing, however, and many hibakusha understandably felt anxious about possible genetic effects.

11. Interview with Kuniko Jenkins, June 5, 1976, San Francisco.

12. Shigeko Sasamori, *Shigeko, Go On* (in Japanese) (Tokyo: Chōbunsha, 1982).

13. Michiko Ishimure, "Kiku to Nagasaki" (The Chrysanthemum and Nagasaki), *Asahi Jaanaru*, August 11, 1968, pp. 4–9.

eight

The Death of the "President's Patient"

There is a tragic story, widely known among hibakusha in southern California, concerning Mary Yano, who was born in Los Angeles and exposed to the bomb in Hiroshima, where she was attending high school. Yano became sick after she returned to the United States and died of cancer after twelve long years of fighting "A-bomb disease."

Mary Yano was believed by many to have written to President Johnson in the course of her illness, asking the president to help her obtain medical aid. According to the story—it achieved something of the status of a myth—Yano's request was granted. Yano must have believed she had a right to request medical care, but it was a very brave move to write such a letter, considering the difficult situation the Kibei Nisei found themselves in after they returned from Japan, the wartime enemy.

Where is the letter Mary Yano supposedly wrote to President Johnson? What happened to it? Did she leave a copy behind somewhere? Her mother, Toshiko, who still clung to her daughter's unused wedding gown, was determined to keep Mary's room undisturbed and her belongings untouched, so it was not possible to search for the letter in the Yano house. Could the president's reply be located, perhaps in the files of the Los Angeles County Hospital, where Yano was treated? "Because of President Johnson's letter," Mrs. Yano claimed, "the medical care was free. But his letter was sent to the hospital. It may still be there."[1]

With the consent of the patient's family, and the assistance of then Los Angeles County Medical Examiner-Coroner Thomas Noguchi, I was able to obtain Yano's medical records.[2] President Johnson's letter was nowhere to be found in that file, but there was a copy of Yano's letter of appeal. Oddly enough, the letter had been retyped on plain white paper without any letterhead and without the name of the addressee. The letter bears the date, "April 22, 1958," which would mean that the president at the time was not Lyndon Johnson, but Dwight D. Eisenhower.

The letter, which begins "Dear Sir" rather than "Dear Mr. President," appealed as follows:[3]

> I am a citizen of the United States and had been in Hiroshima City in the street car when the Atomic bomb was dropped on August 6, 1945. It was dropped about a mile away. I felt the heat of the bomb. When I got off the street car and I looked back, the street car was burning. The dust got in my ears, eyes, nose and mouth. A few days later, although I received no burns, my hair started to fall off. For years, I felt that the Atomic bomb had no effect on me so I had done domestic work, but about four years ago I started to get tired for no reasons, and started getting pains in my legs. I had a preliminary examination at the County Hospital and it is believed that it is due to degeneration of the bone marrow caused by the Atomic bomb. I also have anemia. I am going to have medical treatment at the County Hospital in May. I would like to get the aid of the Federal Public Health as soon as possible. At present, I have difficulty in walking, and am not able to go up or down the stairs without help. I went to Japan in May, 1940, with my mother and brothers to visit our grandparents whom we had never seen. We were unable to come back to the United States because of the outbreak of the war, and the ships were tied up. We were stranded there.
>
> Yours very truly,
>
> Miss Mary Yano

This is a moving document for the very reason that Yano makes no attempt to exploit the emotional possibilities of her situation. Sticking to the facts, she simply asks her government, which had dropped the A-bomb, to supply modest medical aid. Between the lines of her subdued letter, however, you can sense her emotional turmoil.

Mary Yano was born in Los Angeles in 1924. Her Japanese name was Merii, which is pronounced "Mary." Her father was a successful businessman, serving as vice-chairman of the Hiroshima Prefectural Society of Southern California and in other prominent posts. Mary's mother came from a family of twenty-four generations of Buddhist priests in the village of Kabe, Hiroshima Prefecture. In the spring of 1940, Mary, her mother, and four brothers boarded a ship bound for Japan to visit their grandparents. Mary's father remained behind. He planned to have his sons educated in Hiroshima, to instill in them the Japanese spirit, and he probably had no idea that war between the United States and Japan was imminent. The family took practically everything with them—piano, beds, and washing machine included. Toshiko Yano recalled that they had more luggage than anyone else on the ship.[4]

In Hiroshima, the difference in climate took its toll, and the Yano children were constantly sick. They were thus unable to make the long trip home, even when it became clear that war was likely to break out. After the war began, Mr. Yano gave up all his property and sailed for Japan on an exchange ship. He was in Hiroshima when the bomb was dropped.

In the summer of 1945, Mary Yano was still at Hiroshima Jogakuin (a girls' mission school), where her lack of ability in Japanese language delayed her advancement. The A-bomb was dropped while she was riding in a streetcar near Tera-machi. The heat was so intense that a fan in her hand caught fire, but she was not burned. As she sought refuge through the shower of black rain, along with a friend and the friend's mother, Mary was fortunate to be picked up by a truck and brought home to Kabe.

Mary's father and her oldest brother, who had been in the city at the time of the bomb, returned home without injuries. Mr. Yano came back to the United States in 1954 and died in 1958 of pancreatic cancer. The eldest Yano son also became sick and was unable to walk after his return. He was later confined to his home.

Mary worked for a while as an interpreter for the occupation in Kure, but soon returned to the United States. She was healthy at first. She worked hard to help her younger brothers regain their U.S. citizenship. (They had been forced to become Japanese citizens when they entered middle school in Hiroshima.) In 1954, her parents and her eldest brother returned. Although Mr. and Mrs. Yano did not have U.S. citizenship, immigration officers in Hawaii readily gave them landing permits when they explained that two of their sons had served and been injured in the Korean War.

Soon after the family was reunited and started rebuilding the life that had been destroyed by the war, Mary's health began to deteriorate. It was in 1958, after her father died of cancer, that Mary wrote to the government, perhaps to the president. She had already been suffering ill health for four years. She began to have difficulty walking, so she could no longer work. Her savings were soon exhausted by medical bills, and her health insurance was cut off. She must have been desperate by the time she wrote her letter to the government.

Mary's letter eventually reached a high official at the Pentagon, Dr. E. H. Cushing, the deputy assistant secretary of defense for health and medical affairs. He responded in the following letter to Dr. Stafford Warren:[5]

The Death of the "President's Patient" 83

May 15, 1958

Dear Staff:

It is many a day since we have seen each other and I am sorry our contacts have not been maintained. Nonetheless I am taking this opportunity to ask for your help in connection with a letter that reached my desk several days ago. It is really an interesting case of an American citizen who was in Hiroshima when the bomb was dropped and had minor results. Would it be possible for someone on your staff to check this woman over and make sure that she is not as she claims to be, a bomb victim, at the present time? Any word that comes from you would be very helpful.

Sincerely,

E.H. Cushing, M.D.
Deputy Assistant Secretary

Enclosure
Dr. Stafford L. Warren
141 Tigertail Road
Los Angeles, California

It is clear from its tone that the letter is not an official request. It is simply a favor asked of an old friend. But there may have been deeper concerns and meaning behind the casual phrasing. If there were American citizens in Hiroshima when the A-bomb was dropped, the matter of compensation could be brought up. And if the number of victims was not just one or two, Mary Yano's case could establish an important precedent. That is probably why Cushing wished to make sure that she was *not* an A-bomb victim.

Stafford Warren—chief of the medical division of the Manhattan Project survey group, and one of the first Americans to enter Hiroshima after the atomic bombing—had been at the UCLA medical school since 1947, and was dean of the school at the time of Cushing's letter. Warren immediately wrote to Los Angeles County Hospital:[6]

June 12, 1958

Roger O. Egeberg, M.D.
Medical Director
Los Angeles County General Hospital
Los Angeles, California

Dear Dr. Egeberg:

Miss Mary Yano, who was examined at the County Hospital presumably in April or May of this year, has made a petition through the Office of the Assistant Secretary of Defense for assistance.

I would appreciate having a transcript of her medical record so that I may inquire as to the status of United States' citizens caught in the Hiroshima bombing, to see whether the Atomic Energy Commission or some other agency would be interested in any follow-up.

Cordially,

Stafford L. Warren, M.D.
Dean

Los Angeles County Hospital responded on June 20, 1958:[7]

Yano, Mary
P. F. 1837 020

Dear Doctor Warren:

Your letter addressed to Dr. Egeberg has been referred to this office for reply.

Review of Miss Yano's medical record indicates that she registered in the Outpatient Department of this hospital April 17, 1958, complaining of "anemia" diagnosed by an outside doctor. Patient was in Hiroshima at the time of the atomic bomb blast. States she has weakness in both legs for past three or four years.

Seen in Hematology Clinic May 7, 1958. States she was told she is anemic (Hemoglobin 14.8—WBC 9200 today). Impression: No anemia at this time. Advise return in six months for recheck.

Seen in Neurology Clinic 5-8-58. Complains of "weak legs" for past three to four years, gradually becoming worse. Difficulty in climbing stairs. Has no trouble with arms. Examination—negative Brudzinski and Kernig. Voice sounds nasal. Motor—good coordination. Poor strength but full motion at feet and knees bilaterally with decreased tone. Sensation intact. Pain around legs up to thighs.

Diagnostic impression: Progressive muscular or lower motor neuron disease of both legs, etiology to be determined.

This was followed by a supplemental report of March 2, 1959, in which Mary Yano's case was diagnosed as something called "Thomsen's disease," which, according to the report, was "due to myotonia and is not due to lower motor neuron disease. It is congenital in origin."[8]

It is not known how or if these records were transmitted to the Pentagon, or what action was taken. If the "congenital" diagnosis was communicated to the Pentagon, the case would have been closed with the conclusion that there was no connection to the A-bomb.

For Mary Yano, however, the problems were far from resolved. On the contrary, her illness—whatever its cause—grew steadily worse. Not only did she lose the use of her legs, but two and a half years after she was diagnosed with Thomsen's disease, lumps were found in her right breast.

In the Los Angeles County Hospital file on Mary Yano, there was also a summary of her medical record, dated July 31, 1962, that had been sent to the Social Security Administration in Los Angeles.⁹ According to that report, Mary had an operation to remove a tumor from her right breast on December 18, 1961. The diagnosis was "probable fibroadenoma," or a benign tumor. Mary's mother believed that her daughter had breast cancer.

The summary record reported, "last seen 6/27/62 in Neurology Clinic for treatment of myotonia congenita." There is no mention of what happened to Mary after that. According to Mrs. Yano, Mary's health insurance was cut off because she was repeatedly hospitalized and underwent surgery. "My daughter made arrangements herself with the County Hospital and called for an ambulance each time. She used to say 'In this country, you have to push as hard as you can to get anything.'"¹⁰

In November 1963, Lyndon Johnson entered the White House. Meanwhile, Mary's condition slowly but steadily deteriorated. If she ever wrote a letter to President Johnson, it must have been around that time. No record of her letter or any answer from the president or the government exists in the County Hospital file, however, and in all probability Yano's free hospital care was covered by existing welfare programs. Although she continued to tell her mother, "I can get treatments because of President Johnson's letter," this might have been just a matter of pride. Yano believed that she had the right to claim medical assistance from the government as an American citizen who was bombed in Hiroshima during the war. By her reasoning, medical care should not have been just a routine part of social welfare, but rather should have been part of the compensation due a hibakusha who was harmed by her own country's A-bomb. "President Johnson's letter" may have been an illusion, but the belief that she was the "president's patient" appears to have sustained Mary Yano emotionally.

In the spring of 1968, President Johnson announced that he would not run for another term in office. By Christmas, Mary Yano was no longer able to get up by herself. She had become deeply religious, kneeling down to pray by her bed every night. She calmly told Mrs. Yano, "Mother, I think I am dying. Please forgive me for becoming a Christian, though I am the granddaughter of a Buddhist priest. We may pass through different doors into the other world, but I am sure we will meet there again."

There is a photograph of Mary sitting on a sofa by a Christmas tree. It shows black eyes set in a pale white face, rather elongated for a Japanese. In April of the following year, cancer finally invaded her throat and Mary Yano died. As Mrs. Yano recalled, "On New Year's Day, we found tropical fish, which she treasured, dead in the tank, and a mirror that we had bought at the end of the year from Bullock's department store broken in two. The store wouldn't believe it, but they replaced it anyway. Then, the head of a Hakata doll came off when it arrived from Japan. We couldn't help wishing the doll might have fallen in Mary's place."¹¹

And so the president's patient died.

In April 1978, nine years after Mary Yano's death, I had the opportunity of showing some of this correspondence to Dr. Warren. Still healthy and active at the age of 83, he met me at the UCLA Medical Center in a building, Warren Hall, named in his honor. In a small research room, he began reading Mary Yano's record with some deep feeling.

"Yes, I remember the letters I wrote," he said. "I knew Dr. Cushing very well. By the way, this patient was not a real victim of the A-bomb. No, she certainly was not. In no way were her troubles related to the A-bomb."

"But she died of cancer."

"What kind of cancer?"

"Breast cancer, which finally spread to the throat."

"No," Warren declared emphatically. "Her cancer has nothing to do with the A-bomb." He looked at her medical record. "Her blood was normal. Besides, she survived twenty years after the bombing and she did not have any burns or keloids. It was summer, and most women were wearing light clothing with short sleeves, but she was on a train, so she might have been behind somebody. Her location was on the edge of the impact area. So, she was just outside the critical zone where neutrons and the pressure of the blast reached. I remember a train burnt to its frame in Hiroshima. By the way, talking about Dr. Cushing, he was a man with a negative attitude."

"What do you mean?"

"He was the type of man who would not commit himself actively to responsibility in anything. That may be why he ended up at that level. Of course, the deputy assistant secretary of defense is a rather high position. But he could have gone higher. His father was a very famous neurosurgeon, too famous. He remained in his father's shadow. . . . I don't like the words 'make sure that she is not as she claims to be, a bomb victim.' Cushing obviously had a preconception. He did not want to get involved with this patient's case. You see, if her case was determined to be a result of the A-bomb, many other bomb victims would swarm to the Pentagon."

"Does that mean the Pentagon knew there were other American civilians in Hiroshima at the time of the bombing?"

Warren did not answer this question.

"Anyway, her case was mainly neurological in nature. I don't see a particularly high rate of cancer among the survivors. The effect of radiation is mostly in the lymph glands or the destruction of cells, like leukemia. There's a recent theory that cancer may be caused by a virus. Radiation might make a virus more active. Anyway, her case was hereditary, 'congenital in origin' as it says here. There must be somebody with these symptoms in her family—even if it's not her immediate family. You've got a poor case!" Warren declared.

I am not qualified to judge whether Mary Yano's illness was a result of the A-bomb or not. But after being bombed in Hiroshima, she suffered a series of maladies, including breast cancer, until she finally died of cancer of the throat. In the years since my conversation with Warren, it has been established that there is a higher incidence of breast cancer among hibakusha than among the general Japanese population, especially among those who were exposed to the bomb at an early age. However, without a careful determination of the dose of radiation Mary Yano received, it is not possible to estimate the likelihood that her cancer was caused by the bomb.

In any case, Mary Yano died as a hibakusha, believing, to the end, that she was the "president's patient."

Notes

1. Interview with Toshiko Yano, July 1, 1976, Los Angeles.
2. File on Mary Yano (hereafter cited as *MYF*), Los Angeles County Hospital (hereafter cited as LACH). Used with the permission of Toshiko Yano.
3. *MYF*, a retyped copy of Mary Yano's letter to "Dear Sir," dated April 22, 1958.
4. Interview with Yano.
5. *MYF*, E. H. Cushing's letter to Stafford Warren, dated May 15, 1958.
6. *MYF*, Warren's letter to LACH, dated June 12, 1958.
7. *MYF*, LACH's letter to Warren, dated June 20, 1958.
8. *MYF*, the follow-up report from LACH to Warren, March 2, 1959.
9. *MYF*, the medical record on Mary Yano prepared by LACH and sent to the Federal Social Security Administration in Los Angeles, dated July 31, 1962.
10. Interview with Yano.
11. Interview with Yano.

nine

The Hibakusha Begin to Organize

For a Nisei who survived the atomic bomb to assert her rights as a citizen and make a direct appeal to the president of the United States must have taken all the courage she could muster. Mary Yano's plea is the only known instance of this ever happening. Because hers was an unprecedented act, it was still talked about among the hibakusha years later. Yano understood the American character and psychology. She knew that go-getters command respect, and that if you want something you must speak up and be persistent. She had the gumption and courage to put this belief into action. One other factor was to her advantage: She had a command of the English language to go with her courage.

Many survivors shared similar experiences—their formative years had been spent in Japan, and they had returned to the United States as adults—but their strong spirits were often overshadowed by a language handicap. The delayed returnees and those who came to the United States to marry also faced enormous problems. Their sense of helplessness and despair ran deep. Misery loves company, and the hibakusha often banded together to seek solace in their indifferent surroundings.

On August 6, 1965, the following notice appeared in the *Rafu Shimpō* and *Kashū Mainichi*, the two Japanese-language newspapers in Los Angeles:

A Call to All Hiroshima and Nagasaki A-Bomb Survivors
  A meeting will be held to organize a Hibakusha Friendship Group (temporary name). Those who are interested, please contact:
    Shuji Okuno SY 1–1576 or Kaname Shimoda AN 2–3250.[1]

Shimoda, born in Sacramento, had experienced the A-bomb in Hiroshima when he was fourteen and returned to the United States after enlisting in the army at the age of twenty-seven. According to Ernest (Satoru) Arai, a Hawaiian-born Sansei who was also a hibakusha and a friend of Shimoda, the reason for forming a group was to launch a drinking fellowship of kindred souls.[2] (Shuji Okuno subsequently left the group.)

Some hibakusha had come to believe that alcohol could lessen, if not cure, the aftereffects of the atomic bomb. Ernest Arai's father had developed telltale spots on his body and was on the verge of death when he began to drink to excess. He claimed that it was his drinking that cured him, and he was still alive as of the summer of 1976. Whatever the virtues of alcohol may have been, the hibakusha shared common problems, and commiseration was their only consolation in an alien land.

Borrowing Shimoda's expression, August 6 is the day for hibakusha to commemorate the "accident of their survival." The idea behind the group was that fellow survivors of the mushroom cloud would get together and, in despair, drink themselves into a stupor so they could forget the past.[3] It was with this thought in mind that the first meeting was proposed.

In traditional Japanese society, only men engaged in drinking bouts. However, according to Tomoe Okai, it was mainly women, alien wives of U.S. citizens, rather than Kibei Nisei, who phoned in response to the notice in the papers.[4] Okai, whose first husband was killed by the A-bomb, had remarried to a Nisei and in January 1962 had come to live in Los Angeles. Immediately after her arrival, she felt herself growing physically weak, and in September 1962 she returned to Hiroshima for a thorough physical examination and diagnosis. While she was still in Japan, Okai began to suffer from the effects of the A-bomb, and her physician informed her that many of her internal organs were affected. Despite this diagnosis in Japan, American specialists told Okai that she was suffering from anxiety, not from aftereffects of the A-bomb.

When she talked to other troubled hibakusha, Okai learned that she was not alone. Others shared her experience of encountering American doctors who were utterly lacking in knowledge and understanding of the A-bomb syndrome. Needless to say, there was a language barrier, but what made the hibakusha most uneasy was the ignorance and unsympathetic attitude of American doctors. The survivors suffered an array of symptoms: They tired easily; they often fainted; vomiting was frequent; they lost their balance easily; they had nose bleeds and bloody stools that would not stop.

Families had been divided by the war, and one special agony for hibakusha was that the members of their families who had remained in the United States refused to understand their plight. They would look coldly upon a hibakusha relative, and say, "He's complaining again," or "She's complaining again." The children of older hibakusha made little effort to empathize with the unbearable feelings their parents were experiencing.

The first meeting of the Hibakusha Friendship Group was set for August 12, 1965. This date subsequently gained significance in American history because it was the day after a riot broke out in the Watts section of Los Angeles. The riot continued for six days, arson was widespread, and a number of people were killed.

All traffic leading to Watts was stopped and a sense of emergency hovered over the entire city for a week. In Little Tokyo, the annual Nisei Week Festival was in full swing, but the parade that marked the end of the festival had to be canceled because of the riot. The Watts riot may also explain why only six people gathered for the first A-bomb survivors' meeting. The six who came were Kaname Shimoda and Shuji Okuno, the initiators; Shimoda's friend, Ernest Arai; and Tomoe Okai, Shin'ichi Kato, and Chizuko Favatella.

Favatella was married to a white American and had been in Nagasaki when the bomb fell on that city. Kato had been a newspaper reporter in Los Angeles before the war.[5] He was sent to Japan from the detention camp in Missoula, Montana, on the first exchange ship. He returned to Hiroshima, and was working at the *Chūgoku Shimbun,* a newspaper, when the bomb fell. Both his younger brother and sister died from the blast. After the war, he returned to Los Angeles, where he was engaged in newspaper editing and compiling a history of Japanese immigrants to the United States. He served for a time as a vice president of the Hiroshima Prefectural Society (an association of people with roots in Hiroshima Prefecture), but he never revealed to anyone that he had experienced the A-bomb. His rationale was that Japanese Americans "hate to talk about the atomic bomb. They feel they are living here at the sufferance of Americans and they should not mention anything that Americans dislike hearing about." The only reason Kato came to the meeting was that it had been organized as a social gathering among fellow survivors. He had known Okai since her Hiroshima days.

This first meeting of the group was doomed from the start, Watts or no Watts. As Kato pointed out, in the country that caused the atomic disaster, it was an act of the utmost courage for anyone to step forward and announce that he or she was one of the surviving victims. Living evidence of the A-bomb tragedy, in the form of human beings who experienced it, would arouse in the perpetrators that sense of guilt that they have sought to suppress into a remote corner of their consciousness.

The Kibei Nisei were also self-conscious of the fact that they had sat out the war on enemy territory. Cultural differences and the language barrier made it unlikely they would take a stand and attempt to explain their experiences. The majority of those who came to the United States to marry also felt, as Kato expressed it, that they were in the United States by the grace of the American people. This was the same frame of mind that prevailed among the Issei immigrants when they first came to America. It was perhaps a reflection of the continuing bias of white Americans against Asians, even though relations between Japan and the United States after the war had shifted from hostile to friendly in a way that would have been unthinkable in the prewar era.

I encountered this reticence, and additional motivations for it, when I approached the leader of a local chapter of a hibakusha group to ask him to arrange a meeting with A-bomb survivors living in his area. This was his response:

> The summer season is an awfully busy time here, and people are committed to their work. On top of that, the Bon festival is coming up soon, so people are very busy. [When I mentioned the purpose of my research, he responded] If that is the purpose, I'm afraid no one will talk to you. All of them want to forget the past. If for any reason their A-bomb experience becomes known, their insurance would be terminated immediately. In this country, everyone takes out their own medical insurance. You say you're going to write a book about them. Even if the person's name is disguised, he is sure to be found out, and before long he'll get a termination notice from the insurance company.
>
> Without insurance, your entire income would soon go to doctors. You see, these people don't even join the hibakusha association for fear of exposing their past. What you are asking is out of the question. In a big city like Los Angeles, it might be different. The Japanese community here is very small. Children get married and the families become intimate and soon everybody knows [that someone is a survivor]. Small town people are stubborn, once they make up their minds. I know they won't meet with you. Everyone will be against it, so I don't think I'll get together with you either.[6]

Going public with their A-bomb experience would not relieve the trauma branded onto the souls and bodies of the hibakusha. They were convinced that the American government was not going to extend a helping hand and that no one would understand even if they talked about their experiences. So many hibakusha lived lives of quiet desperation and pushed their memories of that day into a dark corner of their consciousness.

The slightest problem with their health, however, caused the hibakusha great anxiety. The trauma of the A-bomb experience was too intense to be ignored and it was inevitable that all health problems would somehow be attributed to the bomb. As long as medical knowledge could not clearly determine the effects of radiation and how long they continue, it was only natural that survivors connected their ailments to the bomb. And they were intensely aware of the high cost of medical care, even though American doctors seemed to have little knowledge of radiation disease.

Insurance companies were insensitive to A-bomb survivors. Though patients' records are supposed to be confidential, it seemed that insurance companies obtained information from doctors. Ailments caused by the A-bomb were often excluded from coverage, or in other cases hibakusha saw their insurance premiums soar. Kaz Suyeishi reported that she received a notice from her insurance company stating that any condition arising from contusion (caused by the bombing) would not be covered. Tokuso Kuramoto's group insurance coverage included a clause that clearly stated, "Conditions arising from act of war or atomic bomb explosion, or radiation from any nuclear sources shall not be covered."[7]

If the hibakusha were to gather together in an association, it would be with the hope that they might obtain sympathetic and effective medical treatment for their bomb-related ailments and psychic stress. At the time of the first hibakusha meeting, twenty years after the atomic inferno, there seemed little reason to entertain any hope for proper medical care in the United States. If medical attention was to be arranged, the only path was to bring specialists from Hiroshima.

At first, only minimal efforts were made to obtain this medical care. In October 1965, a round-table discussion was held between hibakusha and reporters from the Chugoku Broadcasting Company and Hiroshima Television, who were in the Los Angeles area to gather newscast material. The "Tomo no kai" (Friendship Group) was pulled together in time for that meeting, held in the hall of the Southern California Japanese Chamber of Commerce, but it was an organization in name only.

Two years later, in October 1967, Tomoe Okai made a visit to her family home in Hiroshima, and while there, she met with Setsuo Yamada, the mayor of Hiroshima. This represented the first time that American A-bomb survivors had met with officials in the city government. The meeting was arranged by Shin'ichi Kato, one of the six who attended the meeting in August 1965. Kato had since returned to Hiroshima and was living there when Okai visited.[8]

Okai asked Mayor Yamada to send a team of A-bomb disease specialists from Hiroshima to the United States. However, the practice of medicine by foreign doctors, including physical examinations, is prohibited in the United States, just as it is in Japan. Even if funding and capable personnel were available, this legal prohibition would bar a medical mission from Hiroshima. Still, Yamada concluded the meeting by saying that, despite the legal obstacles, he would do whatever he could to make it happen. He promised to make inquiries at the Ministry of Foreign Affairs and the American embassy in Tokyo. The mayor's statement was carried in the *Chūgoku Shimbun* on October 4, 1967, but the mayor had little idea of how difficult a road he was embarking on.

During her time in Hiroshima, Okai also visited the Atomic Bomb Casualty Commission (ABCC) and obtained the names and addresses of physicians and scientists who had previously worked at the commission and were then living in the Los Angeles area. After she returned home, Okai attempted to contact a blood pathologist who had been at the UCLA Medical School, but to her great disappointment, she found that he had moved to Hawaii. She tried to get in touch with others on the list without success. The language barrier and limited transportation made her task all but impossible. She finally decided that it would be more fruitful to seek help from Japan.

With this in mind, she and Kaname Shimoda visited the Japanese consulate in Los Angeles in November 1967. They met with a vice-consul to ask about the possibility of having a medical team from Japan sent to the United States. The vice-consul responded, in the perfunctory manner typical of government officials, that nothing could be done in California; that the memory of Pearl Harbor revived every year on December 7 kept latent hostility alive; and that the visitors should try the Japanese embassy in Washington.

Throughout the thirty-minute interview, this minor official sat with both feet resting on a coffee table. The visitors felt deep anger at this treatment, which went far beyond simple discourtesy and lack of consideration. From the time of the Meiji period, in the latter half of the nineteenth and early twentieth century, Japanese overseas officials had shown contempt for Japanese immigrants. Perhaps this tendency had been inherited by their postwar counterparts. Or perhaps this particular official considered the visitors troublemakers, reminders of the thorny problems that the atomic bombings pose to U.S.-Japan relations. Okai and Shimoda read these messages in the attitude of this official.

The U.S. government was unresponsive to begin with, and when Japanese officials treated them with such indifference, they felt as if they had been thrown into an abyss between the two countries. The only place left to pin their hopes on was the city of Hiroshima. In 1968, Mayor Yamada came to Long Beach, California, to attend a U.S.-Japan mayors' conference, and he stopped over in Los Angeles. Twenty A-bomb survivors met with him at the hall of the Hiroshima Prefectural Society and appealed to him for help. After the mayor's return to Hiroshima he sent a terse note which read, "Your request for a medical team is presently being studied."

After this, Okai and her group continued to approach every VIP from Hiroshima who passed through the city, be they a vice mayor or a Buddhist monk, with their appeal for help. They asked the Southern California Hiroshima Prefectural Society for assistance early on, and four presidents of the organization came and went, but no substantive action was taken. Their only advice was for the group to write to the mayor of Hiroshima. The survivors felt isolated, their efforts seemed futile. Inaction continued.

In 1971 there was finally a breakthrough. In June, Okai, the leader of the survivors' group, made a fourth visit to Japan to get treatment of her ailments. During her stay she appeared on a program of the national NHK television network and made a heart-rending appeal on the plight of the American hibakusha. Since 1965, Hiroshima television stations had occasionally presented appeals for aid to the U.S. hibakusha, but these efforts had been sporadic and ineffectual. The broadcast over NHK was the first time a nationwide TV appeal had been made.

Okai wanted to meet with Mayor Yamada again, but he was preoccupied with a reelection campaign and sent word that he was unable to see her. A meeting was eventually arranged, however, through an unexpected channel, the World Friendship Center in the Midori-machi district of Hiroshima. The WFC was set up to appeal for world peace by providing assistance to A-bomb survivors. It so happened that the center had invited several like-minded American clergymen and private citizens to Japan, and at a welcoming banquet on June 6, Dr. Tōmin Harada announced that the WFC hoped to solicit contributions throughout Japan to send a team of medical specialists to the United States. The following morning a report on this plan was carried in the *Chūgoku Shimbun*, and Mayor Yamada immediately sent a note to Okai telling her that he wanted to meet with her.

On June 9, Okai, Harada, and Kato met with the mayor. Representatives from the ABCC and the two survivors also came together in a subsequent meeting. George Darling, director of the ABCC, and Hiroshi Maki, deputy director, offered their whole-hearted support to the effort. Dr. LeRoy R. Allen, special assistant to the director, happened to be in Washington at the time and he was to be asked to sound out how the U.S. government would react to this plan.

Okai returned to California in mid-August. She had been asked by NHK to solicit the assistance of hibakusha in Los Angeles in the network's effort to reconstruct a map of the center of Hiroshima to show how the city had been before the A-bomb. Okai went to see Sumiko Tatematsu of Asahi Homecast, a Japanese-language broadcasting group, to enlist her help in publicizing the NHK project. This was the first time that Okai had met Tatematsu, who turned out to be very sympathetic to the problems of hibakusha. Her advice to Okai was to go see Thomas Noguchi, Los Angeles County chief medical examiner-coroner.[9]

Noguchi was already well-known throughout the country. He had come to the United States after the war, having graduated from a Japanese medical college. He overcame the handicaps facing a newcomer, and by sheer hard work and ability had risen to the position of coroner of Los Angeles County, whose extensive bailiwick is the largest of any county in the nation. He had the responsibility to ascertain the cause of any death that occurred in metropolitan Los Angeles. The sudden death of Marilyn Monroe, the assassination of Senator Robert Kennedy, and the sensational Sharon Tate murder case all came under his forensic jurisdiction.

As a postwar immigrant who had risen to one of the most powerful positions in greater Los Angeles, Noguchi soon became a target of jealousy and insidious personal attack. The animosity finally took its toll, and he was relieved of his duties. But Noguchi was made of stern stuff, and he fought back with determination. A public hearing was held, with full radio and TV coverage, and his case became a cause célèbre. The line that had long separated American-born Japanese Americans and postwar immigrants completely disappeared, and the L.A. Japanese community stood unanimous in Noguchi's support. A support committee was formed that quickly raised $50,000 to fund his legal battle. In the summer of 1969, Noguchi was vindicated.[10]

Tatematsu knew that Noguchi felt a sense of obligation and that he wanted to do some good for the community in return for the support it had given him. Aid to the hibakusha would be very much in his province, since it involved medical knowledge of a very special kind. Tatematsu's intuition was correct: Noguchi readily responded to the request to help the survivors, and he acted swiftly and pointedly. He advised the group to dissolve the ineffectual Friendship Group and to assemble an organization with real backbone.

On October 13, 1971, at a Chinese restaurant in Little Tokyo, an organization was formed called the Committee of Atomic Bomb Survivors in the United States (CABSUS). The officers were Tomoe Okai, president; Ernest Arai, vice president; and Kaname Shimoda, secretary. A total of thirty hibakusha came to the meeting. The Hiroshima Prefectural Society of Southern California for the first time voiced support for the group.

While these developments were occurring in Los Angeles, propitious news came from Hiroshima. The mayor had received an encouraging report from Dr. Allen of the ABCC, who had returned to Hiroshima after completing his work in Washington. This news appeared in the October 16 *Mainichi Shimbun* as follows:[11]

> Dr. Allen of the ABCC . . . arrived at the conclusion that, although Japanese doctors would not be allowed to practice medicine in the United States, if the mayors of the several U.S. cities involved entered into an agreement with the mayor of Hiroshima to allow Japanese medical teams to visit the cities for research purposes, the visit could be arranged.

On the basis of this news, Mayor Yamada announced that he would meet with the mayors of the involved cities at the U.S.-Japan Mayors Conference, to be held in Kyoto starting October 20, and obtain their permission for a visit by a Japanese medical research team. Yamada went so far as to say that he would not mind leading the team himself, according to the *Chūgoku Shimbun*.[12] To make such confident statements, Yamada must have been in communication with Los Angeles Mayor Sam Yorty, who was scheduled to attend the mayors conference.

The news of the possibility of a visit by a Japanese medical research team was a godsend for the U.S. survivors' group. Okai declared: "Our efforts are now bearing fruit, thanks to the people of Japan."

But it turned out to be too early to revel in this good news. It was not until six years later, in March 1977, that a team of doctors actually came to the United States from Hiroshima.

Notes

1. *Rafu Shimpō* and *Kashū Mainichi,* August 6 and 7, 1965.
2. Telephone interview with Ernest (Satoru) Arai, June 19, 1976, Los Angeles.
3. Interview with Kaname Shimoda.
4. Interviews with Tomoe Okai, June 17, 1976, Los Angeles; and January 21, 1978, Tokyo.
5. Interview with Shin'ichi Kato, March 4, 1977, Hiroshima.
6. Telephone interview with an anonymous survivor, June 13, 1976, San Jose, California.
7. Interview with Tokuso Kuramoto, October 23, 1976, Los Angeles.
8. Based on interviews with Okai and Kato.
9. Interview with Sumiko Tatematsu, June 18, 1976, Los Angeles.
10. On the case of Thomas Noguchi, see Masakiyo Watanabe, *Misshon rōdo* (Mission road) (Tokyo: Ushio Shuppansha, 1991).
11. *Mainichi Shimbun,* October 16, 1971.
12. *Chūgoku Shimbun,* October 16, 1971.

ten

Hibakusha Discovered

The news that doctors were coming from Hiroshima cheered hibakusha in the United States, who had previously experienced only frustration in obtaining medical assistance from both the U.S. government and from Japan. Now at last, Hiroshima, which was for many of them their home town, was extending a helping hand.

Having achieved a certain official recognition, the hibakusha began to come forward and to reexamine themselves and their history. A meeting was organized, with the help of Dr. Noguchi, to discuss the hibakusha's problems from the perspective of medical specialists. The meeting took place on November 3, 1971, at the Japanese Chamber of Commerce building in the Little Tokyo district of downtown Los Angeles.

Dr. Jack Kirschbaum of Palm Springs, a hematologist and a former researcher with the Nagasaki ABCC, addressed the gathering on the effects of radiation on the human body. About one hundred people listened to his lecture, in which he cautioned that not all of the symptoms suffered by hibakusha were related to the atomic bomb. Many who attended were not hibakusha, among them George Takei, a Japanese-American actor who was active in the anti-nuclear movement.[1]

This meeting broke the ice. Gradually, the hibakusha, who had been silent for so long, began to speak out beyond the confines of the Japanese American communities, and to address the entire United States.

In early April 1972, *Newsweek* ran a full-page article on the woes of American hibakusha.[2] The article was accompanied by photographs of two hibakusha, Kaname Shimoda of Pasadena and his friend, Ernest Arai. These two expressed the common complaints that American doctors lacked knowledge about atomic bomb disease and treated it with indifference; that insurance companies raised their premiums and limited their benefits when they found out that a client was an A-bomb victim. The article also described marital troubles and employment problems that caused the hibakusha anxiety. The hibakusha who were interviewed were cautious, afraid that they might inflame the feelings of the American people, but they indicated that they could no longer keep silent.

The *Newsweek* article gave the hibakusha prominent exposure, but it was of flawed value to their cause, because it contained substantial errors.

The article began:

> They call themselves the American hibakusha—literally, the "receive bomb people"—and for more than 26 years, they have lived lives of quiet desperation. While a chastened U.S. government lavished free medical care on survivors of Hiroshima and Nagasaki, the American hibakusha—Japanese too, but living in the U.S.—remained unnoticed and uncared for.

The U.S. government has never been "chastened" in regard to the dropping of the atomic bomb, and it has never "lavished free medical care" on the hibakusha in Japan. The Hiroshima A-Bomb Hospital in Sendamachi, Hiroshima, was built in September 1956 entirely under the auspices of the Japan Red Cross Society. Its construction budget of 70 million yen (about $200,000 at the time) was raised entirely by the sale of New Year's postal lottery cards.[3] The hospital was operated privately, often in the red, and the U.S. government never provided support. (In 1963, the Japanese government began subsidizing its operation.) Other hospitals caring for A-bomb patients have been in the same situation. The *Newsweek* report may have been referring to the ABCC in Hiroshima and Nagasaki, but the ABCC was exclusively a research institution and has never provided medical treatment to hibakusha in Japan.

The myth that the U.S. government had budgeted a large sum every year for the treatment of Japanese atomic bomb victims was widely believed among the American people. It would come as a surprise to many well-intentioned Americans to learn that the ABCC conducts examinations in order to determine the effects of radiation on the human body, but provides no medical treatment. As a matter of fact, the ABCC has long had a reputation in Japan for treating hibakusha as guinea pigs. The U.S. Atomic Energy Commission oversaw the ABCC and provided its operating budget until April 1975, when it was reorganized as the Radiation Effects Research Foundation (RERF), a joint U.S.-Japanese operation. In 1977, the Department of Energy replaced the AEC as the agency responsible for the American portion of the RERF's funding.

Hibakusha Discovered

It stood to reason that American hibakusha wanted an opportunity to consult with doctors who had experience in treating A-bomb patients, and although the *Newsweek* report repeated incorrect information, the very nature of its garbled report gave a boost to the hibakusha in the United States. Most American taxpayers could readily agree that it would indeed be strange to deny medical treatment to American hibakusha of Japanese descent while the U.S. government "lavished free medical care" on Japanese survivors.

At long last, the American hibakusha began to take their case to the whole nation. Soon after the *Newsweek* article appeared, on April 17, the Los Angeles *Herald-Examiner* ran a feature story, headlined "A-Scars Persist from Hiroshima," with pictures of a wasted Hiroshima and of Kaname Shimoda.[4] The story reported in detail Shimoda's A-bomb experience and described the complex problems facing the American hibakusha. It told of how they were speaking out to get medical assistance and of Noguchi's committed efforts on their behalf.

In the interview, Shimoda explained that he still suffered from stomach pains, diarrhea, and prostration. American doctors had not been very helpful, he pointed out, because they had never treated atomic diseases. He also complained that he spent over $1,000 every year in fruitless medical expenses.

"I want something to be done," he pleaded. "The doctors say they know nothing. When we write to people, they don't reply. The government won't listen. Nobody helps us and we need someone."

The *Herald-Examiner* reported that help was finally on the way: "The city of Hiroshima is sending a team of Japanese doctors here and our National Academy of Sciences in Washington is backing the visit."

The mass media had begun to pick up the story of the American hibakusha and make it known to the whole nation. But strangely enough, it was through a quite fortuitous event that one of the highest ranking government officials involved with atomic affairs learned about the American hibakusha.

William O. Doub, who was then a member of the Atomic Energy Commission, was on an official flight from Washington to Chicago in the latter part of April 1972.[5] Casting his eyes around for something to read, Doub borrowed a copy of the *National Enquirer* from a passenger sitting next to him. As he leafed through its pages, Doub happened to come across a picture of an Asian man. The caption announced in large print, "Americans Injured by A-Bombs in Japan Are Refused U.S. Aid Given to Japanese."[6]

This was news to Doub. Although he was a member of AEC and was familiar with the ABCC, he had never known that Americans were living in Hiroshima at the time of the atomic blast.

The man pictured was, once again, Kaname Shimoda, who was identified as the forty-year-old owner of a lawn mower repair shop. His words were very persuasive. Because he was an American, Shimoda explained, he was not eligible for any of the several million dollars being spent to treat Japanese survivors of the bomb. He thought he would get help when he entered the U.S. Army in 1959 and served in Korea. "But the army doctors only said, 'Forget it, Private Shimoda,'" he recalled. "How can I forget it when my stomach hurts?" he asked. Shimoda explained, "We don't have the money to get proper treatment for ourselves. That's why I started this shop: to get more money for treatment." The *National Enquirer* added that there were more than five hundred survivors like Shimoda in the United States.

Reading this article impressed Doub. He told himself, "We have humanitarian responsibilities to all survivors." The article convinced Doub that the government should conduct an immediate investigation and confirm this report of five hundred hibakusha, and on his return to Washington, he promptly proposed to the AEC an investigation of all American hibakusha. Doub's proposal did not initially get majority support on the AEC (the vote was two in favor and three opposed), but within six months the recommendation was approved and arrangements for the investigation were begun in 1973.[7]

By this account, it would appear that Doub had discovered the American hibakusha on his own, but there is more to the story. The staff of the AEC was already aware of the existence of hibakusha. The AEC had been informed of the formation in December 1971 of the Committee of Atomic Bomb Survivors, and after the *Newsweek* article appeared, information concerning American hibakusha had been forwarded to Washington.

"Mr. Doub might have discovered hibakusha in the *National Enquirer*, but our medical department was already investigating this matter," recalled Sidney Marks, a scientist then working for the AEC. "Since we knew of this large hibakusha population, we thought the proper thing to do was to investigate the special nature of this group."[8]

It appears likely that the plan to research the American hibakusha originated in the medical department of the AEC and was subsequently approved by the commission. A national investigation was launched in November 1974, but many winding roads lay ahead before it was put on the right track.

✍ ✍ ✍

The hibakusha broke their long silence because they were encouraged by the good news that a team of doctors from Hiroshima was on the way.

They responded positively to the mass media because they felt assured that they would not remain forsaken.

Media attention provided validation to the hibakusha, but they also knew that disclosure of their A-bomb experience would carry consequences. In Shimoda's case, the premiums for his medical and hospitalization insurance were doubled following the reports in the news media. Still, the situation of the hibakusha could not be improved until they were widely recognized.

One of the first major reports on the American hibakusha in the Japanese press appeared in September 1971, when the *Chūgoku Shimbun,* the Hiroshima newspaper, ran a feature under the headline, "Specialists Are Needed! Misery Under High Medical Expense!"[9] The paper printed interviews with Tomoe Okai and Kaname Shimoda. The story included an unattributed quote that was quite sensational, reporting that the Committee of A-Bomb Survivors was organized "because we could not just watch survivors dying one after another without satisfactory treatment." The exaggerated picture of dying survivors probably was the result of the excitement the hibakusha felt at being able to have their appeal heard in the Japanese media. Shimoda commented later that the expression, "dying one after another," referred to his friends in Japan.[10]

How did American journalists learn of the existence of the hibakusha? Hibakusha were not likely to have approached reporters outside of the Japanese community. Media attention would not have focused on the hibakusha had there not been someone who knew how to bridge the gap between the Nikkei community and the American people in general. The man who fulfilled that role was Dr. Thomas Noguchi. On March 23, 1972, a month before the news media took up the case of the hibakusha, Noguchi submitted a resolution to the California Medical Association:[11]

> WHEREAS, There are approximately 500 citizens of the United States who are survivors of the atomic bombing of Hiroshima and Nagasaki; and
> WHEREAS, Approximately 250 of those are residents of California; and
> WHEREAS, These survivors should have medical evaluation similar to that which is being accomplished by the Atomic Bomb Casualty Commission in Hiroshima and Nagasaki; now, therefore be it
> RESOLVED, That the California Medical Association inform the California Legislature of the plight of those California citizens who are survivors of the atomic bombing of Hiroshima and Nagasaki; and be it further
> RESOLVED, That this problem be referred to the Scientific Board and to the Commission on Legislation for appropriate action.

The resolution was unanimously adopted. Its passage not only informed physicians in California of the existence of the hibakusha, thereby stirring the interest of the general public, but it also prepared the medical profession to welcome a team of doctors from Japan. The most significant step, however, was Noguchi's insistence that the resolution be put before the state legislature, because he realized that the problems of the hibakusha could not be resolved without legislative action.

With this, the hibakusha in California began a long and arduous process of dealing with the state legislature to enact a law providing medical assistance for A-bomb survivors.

Noguchi made another trip to Japan in early April 1972. He met with Mayor Yamada in Hiroshima on April 13 and discussed quite extensively the matter of receiving a group of Japanese medical experts. On April 4 the mayor had announced publicly that, as a result of talks between the ABCC and the Hiroshima University Atomic Radiation Research Center, three physicians, "chiefly from the clinical field," were being considered and that "the city of Hiroshima would absorb the travel expenses of the group and assist in the expense of staying in the U.S."[12]

The purpose of Noguchi's visit to Hiroshima was to help organize the group of doctors and finalize its itinerary. At the meeting on April 13 were two representatives of the ABCC, Dr. Hiroshi Maki, assistant director, and Dr. Leroy R. Allen, associate director; Dr. Fumio Shigetō of the Hiroshima A-Bomb Hospital; and Dr. Tōmin Harada. The meeting resulted in an agreement to send a group of medical experts representing the ABCC, the Hiroshima University Atomic Radiation Research Center, and the Atomic Bomb Hospital, under the title of "survey team" or "consolation team."[13]

The name "survey" or "consolation" was suggested because of the prohibition on foreign physicians performing consultations or treatment, but at the same time, talks were going on with the California Medical Association to permit the group to at least do practical "medical examinations." The departure date was postponed from July to August to provide sufficient time for preparation. Announcements were made that the medical examinations would be held in three cities: Los Angeles, San Francisco, and Honolulu.

Noguchi left for the United States on April 16. In his briefcase he carried a copy of Japan's Act to Provide Medical Treatment for A-Bomb Survivors and the Special Act for Atomic Bomb Casualties. Sending a group of doctors was, after all, a temporary expedient that could not be institutionalized. What could be done to provide proper medical treatment for American hibakusha under the current medical system in the United States? How could the high cost of medical fees be avoided? After the doctors' visit was completed, Noguchi's next goal would be to see to the enactment of a "Hibakusha Medical Assistance Act."

But the visit of the doctors, anticipated with such high hopes by the survivors—who were waiting, in the words of Tomoe Okai, "as if each day is a thousand years"—never took place.

"Sending of Doctors Abandoned," reported the *Chūgoku Shimbun* on July 18, 1972.[14] What happened? The newspaper reported that the plan was abandoned because of American restrictions on medical examinations by foreign doctors, and difficulties the survey team faced in selecting medical experts.

It had been agreed, however, that the medical delegation from Hiroshima would be able to carry out the medical examinations, and the representatives from the ABCC, Hiroshima University and the A-Bomb Hospital had agreed to cooperate on selection of the team. There must have been other complications that came into play. Mayor Yamada admitted to the press, "I never thought it would be so difficult after four years of planning to send a group of doctors."

Dr. Allen, then associate director of the Hiroshima ABCC, recalled that the basic reason for the difficulties was financial. Yamada could not obtain the funding from the city's budget, so he had asked the ABCC for support through Allen. But their assistance was out of the question, since the ABCC was going through severe budget cutbacks at the time. Allen interceded with AEC, but they did not have enough in their budget to invite the medical delegation from Japan. In Japan an effort was made to get aid from the National Preventive Health Research Center, but in vain. Funds to send the medical delegation were nowhere to be found.[15]

Yamada was concerned that something be done to alleviate the disappointment of the hibakusha, and the city of Hiroshima decided to send ABCC Assistant Director Maki to Los Angeles for a consolation visit. Maki was planning to attend an academic conference in Rio de Janeiro in late July, and on his way home he would stop in Los Angeles, meet and talk with the hibakusha, and give them advice about their health management. A five-day visit was set to begin August 5.

Extensive preparations were made on the U.S. side to facilitate the visit. The Los Angeles County Board of Supervisors sent a formal invitation to Maki, stating that if it was necessary to conduct medical examinations and counseling, the facilities of the County Hospital, the coroner's office and other county buildings would be available for his use. This letter was obviously a result of Noguchi's efforts.

Maki arrived in Los Angeles on August 5, the day before the anniversary of the atomic bombing of Hiroshima. On the anniversary every year a memorial service for the A-bomb victims was observed at Nishi-Hongwanji Temple in Little Tokyo. Because of the large number of immigrants with roots in Hiroshima, many who attend the memorial are related to A-bomb victims. At the time of Maki's visit, however, an unusually large number of hibakusha attended the ceremony.

A room in the basement of the temple was set aside that day for hibakusha to register to be examined by Dr. Maki, beginning the following day at the County Hospital. The officers of the Committee of A-Bomb Survivors were astonished by the number of applicants who showed up. Many of those who registered had never come to previous meetings, and they wondered if they would still be eligible to be examined. They had waited for this for a long time, and now their hopes were about to be realized. After a certain amount of confusion, a list of forty-three people was drawn up. With 250 hibakusha estimated to be in California, forty-three in the Los Angeles area alone was quite a sizable number.[16]

Maki served as witness while physicians on the staff of the County Hospital and the University of Southern California Medical School conducted the clinical examinations of the forty-three survivors, complete with blood tests, chest X-rays, and urinalysis. It took the whole day of August 7, but hibakusha and their families waited patiently throughout the day.

A senior county officer who observed the clinic later testified at a state senate hearing:

> I wish to turn away from the somewhat formalized report for just a moment to make a personal observation of that clinic: The survivors could be termed excitedly expectant at the thought that, finally, someone was going to do something to aid them. They came early in the morning, often with their families, friends and supporters to stay the entire day, and without the complaining that often accompanies a day in a similar clinicalized situation we others might know and deal with. They came prepared. They brought toys for the children, boxes of food and drink, and plenty of change for the candy and drink machines in the hospital lobby. They were fully ready to do their part in whatever was needed to prove themselves worthy of aid. They were poked, needled, prodded, dressed and undressed, X-rayed, photographed and interviewed in English and Japanese and asked to remember in detail some very physically and mentally painful and trying years. They gave not one complaint even though the turbulence of this clinic which ran far from as smoothly as might have been hoped; after all, this was the first such clinic of its kind and it was to be a culmination of much hoping and praying by the hibakusha.[17]

Dr. Noguchi later deemed the clinic a "most successful medical evaluation."[18] Dr. Maki may not have anticipated such complete examinations. After all, the official purpose of his visit was to "console" hibakusha. But did this clinic in fact prove to be a real consolation to them? When he returned to Japan, Maki delivered a report to a meeting in Hiroshima City Hall, where he announced that "except for a woman who had a neurological problem, none of the patients would qualify under Japan's A-Bomb medical act."[19]

In order to qualify for medical treatment under the act, survivors must be recognized by the Ministry of Health and Welfare as suffering from diseases caused by the A-bomb blast. The recognized diseases are categorized as follows:[20]

1. Hematopocetic organs: anemia, polycythmia, leukocytopenia, leukocytosis, thrombocytopenia, bleeding tendency, leukemia, erythrocythemia, vera, panmyelocytopenia.

2. Endocrine: thyroid gland, adrenal cortex, dysfunction of sex organs.
3. Malignant neoplasia: cancer, sarcoma.
4. Optical organ: cataract.
5. Liver: chronic liver failure, or cirrhosis.
6. Others: keloid, motor disturbance secondary to burns, and other trauma.

The Japanese law provides qualifying survivors with financial assistance for medical treatment, hospitalization, prescription drugs, surgery and medical examinations, regardless of nationality. But even if the forty-three hibakusha examined in Los Angeles had been in Japan, almost none would have qualified as patients eligible for medical treatment under the Japanese law.

It must have been good news for the hibakusha that the physical effects of their A-bomb exposure were not extensive. But it was too early to conclude that Maki's diagnosis would immediately ease their long years of suffering and anxiety. Even though they were not experiencing A-bomb related ailments at that time, what assurance did they have for their future? The hibakusha knew very well that modern medicine had not found remedies for the various illnesses caused by the atomic bomb. Subconsciously they feared that they might become one of those patients. Because of these fears, many of them sought medical care when even the slightest symptoms appeared. Maki pointed out the problem they faced: "The anxiety of the hibakusha in the U.S. is due to the high cost of medical care—for instance, it costs $130 for the first visit to a doctor—and to the language handicap, which prevents a patient from communicating his subjective symptoms to an American doctor."[21]

Although Maki concluded that only one patient met the Japanese standards for A-bomb disease, Noguchi reported that thirteen of the forty-three had physical abnormalities and eight of them required medical care from a public health perspective. All forty-three felt they needed medical assistance. Since the hibakusha would continue to go to physicians as often as they felt it necessary, the provision of medical assistance remained urgent.

In addition to the California state legislature, Noguchi also targeted Congress for legislation to provide medical assistance to the hibakusha. On August 8, the day after the Maki clinic, the Los Angeles County Board of Supervisors passed a resolution regarding the hibakusha. The county board had no Japanese American members, but Supervisor Warren Dorn was sympathetic and made a motion (obviously at Noguchi's suggestion) as follows: "The Los Angeles County Board of Supervisors wishes to acquaint the President of the United States and the Los Angeles County delegation to the Congress with the problems of the Japanese migrating to this country, and others who are in need of care arising out of atomic bomb injuries, and also to urge an Amendment of Treaty to permit them to continue to receive health aid provided by the Treaty for those remaining in Japan."[22]

The resolution also indicated that hibakusha in Japan were receiving adequate medical care, and those hibakusha who came to this country should be guaranteed similar care. With the help of a member of the staff of Congressman Edward Roybal, a representative from the East Los Angeles area, Noguchi soon came up with a draft of a bill.

Edward Roybal was not only a veteran congressman but he was liberal enough to introduce a bill as unpopular as one that would provide medical assistance to the U.S. atomic bomb survivors. He himself was a Mexican American and was a member of the important House Budget Committee. Many minorities, including Japanese, lived in his district, California's 13th.[23]

The first to join as cosponsors of this bill were Evonne Burke of California and Shirley Chisholm of New York. They were followed by George Brown, also of California, who was known as an outspoken liberal Democrat. It seems that in order to understand the significance of this bill and to be willing to become a cosponsor, one had either to be quite liberal or the representative of an ethnic minority.

In 1972, a bill (HR 17112) to provide medical aid to the hibakusha in the United States was introduced in the House of Representatives, but it died because the election campaign soon began. On January 24, 1973, at the beginning of the 93rd Congress, it was reintroduced as HR 2894:[24]

> To provide reimbursement to individuals for medical relief for physical injury suffered by them that is directly attributable to the explosions of the atomic bombs on Hiroshima and Nagasaki, Japan, in August 1945, and the radioactive fallout from those explosions.
>
> *Be it enacted by the Senate and the House of Representatives of the United States of America in Congress assembled* that (a) the Secretary of the Treasury shall make payments out of any money in the treasury not otherwise appropriated, to any citizen of the United States or any individual who has been admitted to the United States for permanent residence who establishes to the satisfaction of the Director of Office of Emergency Preparedness that he suffered a physical injury that is directly attributable (1) to the explosions of either of the two atomic bombs dropped by the United States on Hiroshima and Nagasaki, Japan in August 1945 or (2) to the radioactive fallout from those explosions. Such payments shall cover expenses incurred after the date of enactment of this Act in the United States, as defined by section 102 (2) of the Disaster Relief Act of 1970 (42 U.S.C. 4402 (2)), by such citizen or individual for the remedial treatment of such injury, including hospital, doctor, and similar medical expenses.

The bill was introduced as an amendment to the Disaster Relief Act and clearly provided only reimbursement for medical expenses after the passage of the act. It did not provide reimbursement for past expenditures, or compensation for the loss U.S. citizens suffered by the bomb. In short, it was a weak bill.

Yet even this modest bill had a hard time getting consideration in the House of Representatives. Because it involved compensation from the federal government, HR 2894 was referred to the House Judiciary Committee. Unfortunately, the whole 93rd Congress was in turmoil at the time because of the Watergate scandals. The House Judiciary Committee was especially preoccupied with deliberations concerning the impeachment of President Nixon. As a result, the bill died.

Even without Watergate, it would have been very difficult to push any bill like this through Congress. Washington was farther away than the hibakusha had ever supposed. Even as they kept hoping for some progress at the federal level, Noguchi and the hibakusha were pursuing the alternate strategy, to pass a bill at the state level. For this purpose, they had to enlarge and strengthen the hibakusha organization. Four new chapters were established, in San Francisco, East (San Francisco) Bay, San Jose, and Sacramento. These chapters were organized into the Northern California Committee of Atomic Bomb Survivors in the U.S. In March 1974, Kanji Kuramoto became the president. There were estimated to be 150 hibakusha in northern California, and the names and addresses of eighty individuals were confirmed.[25]

Having expanded their organization, the hibakusha in California succeeded in getting a hearing in the State Senate before the Subcommittee on Medical Education and Health Needs, a prerequisite to the finalization of a bill. Chairman Mervyn Dymally, in his opening remarks, emphasized the peculiarity of the problem faced by the hibakusha in the United States:

> The Americans who found themselves caught in their parents' homeland at the time of the blast are not eligible for Japanese aid and there is no program in the United States to aid them. Whatever one may think about their military necessity, the only two atomic bomb attacks in the world were in Japan and they were tragic. We have now added irony to tragedy in that the native-born Americans who bore those attacks are the only survivors for whom no program or treatment has been provided.[26]

The existence of the hibakusha, or "bomb-exposed people," is indeed unique. Although there are many who fall between the two countries, Japan and the United States, the hibakusha, in addition to being related by blood and birth to two countries, are pierced through by the atomic bomb tragedy, which is the only one of its kind in human history. They are literally trapped between two countries. At this point, however, even the hibakusha themselves did not fully understand what this meant. The hibakusha, finally "discovered" by the world around them, were also beginning to discover themselves.

Notes

1. *Kashū Mainichi,* November 4, 1971.
2. *Newsweek,* April 10, 1972.
3. On the history of the Hiroshima Atomic Bomb Hospital, see Yoshimasa Matsuzaka and Hiroshima Gembaku Shōgai Taisaku Kyōgikai (Council on Measures for Casualties of the Atomic Bombing of Hiroshima), eds., *Hibakusha to tomo ni: Zoku Hiroshima gembaku iryōshi* (With the Survivors: History of the Medical Treatment of Atomic Bomb Casualties—A Sequel) (Hiroshima: Zaidanhojin Hiroshima Gembaku Shōgai Taisaku Kyōgikai, 1969), pp. 175–177.
4. Los Angeles *Herald-Examiner,* April 17, 1971.
5. The AEC had been organized in 1946 to control atomic energy policy in the United States. It had authority over the whole nuclear system: It was both the developer of nuclear weapons and the promoter of nuclear energy. Doub was one of five members of the commission.
6. *National Enquirer,* April 28, 1972.
7. Telephone interview with William O. Doub, September 8, 1977, Washington, D.C.
8. Telephone interview with Dr. Sydney Marks, April 28, 1978. Richland, WA.
9. *Chūgoku Shimbun,* September 11, 1971.
10. Interview with Shimoda.
11. The text of the resolution is in *Health Problems of Atomic Bomb Survivors: Hearing of the Senate Subcommittee on Medical Education and Health Needs* (hereafter, State Senate Hearing), Los Angeles, May 4, 1974, pp. 52–53.
12. *Chūgoku Shimbun,* April 5, 1972.
13. *Ibid.*
14. *Chūgoku Shimbun,* July 18, 1972.
15. Interview with Dr. Leroy Allen, March 7, 1977, Hiroshima.
16. Interview with Kaz Suyeishi.
17. State Senate Hearing, pp. 42–43.
18. State Senate Hearing, pp. 50–58.
19. *Yomiuri Shimbun,* August 16, 1972.
20. Matsuzaka, *Hibakusha to tomo ni,* p. 155.
21. *Yomiuri Shimbun,* April 5, 1972.
22. Supervisor Dorn's resolution was summarized by Noguchi in the State Senate Hearing, p. 54.
23. On Congressman Roybal and other cosponsors, see Michael Barone, Grant Ujifusa, and Douglas Matthews, *The Almanac of American Politics: 1978* (New York: E.P. Dutton, 1977).
24. The text of HR 2894 was provided to the author by the courtesy of Kanji Kuramoto from his personal papers.
25. The estimated number of hibakusha was supplied by Kuramoto.
26. State Senate Hearings, pp. 1–2.

eleven

You Were Our Enemies!

A public hearing was held on the hibakusha aid bill in Los Angeles on May 4, 1974, before the California State Senate Subcommittee on Medical Education and Health Needs.[1] This represented the most important step in the quest of the U.S. hibakusha for recognition by American authorities. It was the first opportunity for the survivors to express publicly their agony and desperation after nearly thirty years of quiet hopelessness.

Five hibakusha took the witness stand to recount their unforgettable memories of the bomb and describe their continuing travails. Most of them attempted to suppress their emotions and testify as calmly as possible. Two of the five testified in Japanese. Dr. Thomas Noguchi, Los Angeles County Chief Medical Examiner-Coroner; Dr. Sidney Marks, associated with the AEC; and Dr. Jack Kirschbaum, formerly with the Nagasaki ABCC, also testified, along with various state officials. Because many of the hibakusha and their supporters who attended the hearing did not understand English, a volunteer interpreter translated the testimony into Japanese.

Senator Mervyn Dymally, chairman of the subcommittee, was the presiding officer. Senator Alfred H. Song, a member of the subcommittee, was also present with "utmost concern." Song was of Korean descent and represented the Monterey Park district where many of his constituents were Japanese Americans. He was a senior senator, occupying the important position of chairman of the State Senate Judicial Committee.

The first witness was Ernest Arai:

I was born on January 1935, in Honolulu, Hawaii; my parents were also born in Hawaii. In 1940, my father had to leave for Japan and we all went with him. Later we came to the United States.

I was ten years old, and I will never forget the atom bomb in Hiroshima. I was about two miles away from the center when I heard an airplane. I looked for it; a little boy's curiosity. I saw three dots in the sky, then suddenly I saw one explode and a few seconds later I heard a loud noise. I don't know how long I was unconscious, but I awoke in the dark. When I got home I discovered that half of my face, my left arm, my right hand, and my right leg were burned. My mother had to nurse me for four months. The next spring, a GI doctor examined me and found my white blood corpuscle count was only five hundred.

For a couple of years, I felt very weak and was prone to fainting spells. At twenty-one, I came to the United States and went to school for two years, then I started gardening, but I still felt weak. I could work only a couple of hours and had to rest for a couple of hours. I could only work half as much as an average person. About five years ago, my brother opened a service station, so I became an auto mechanic. I changed jobs as this job was easier for me than gardening.

One day I mentioned to one of my customers that we survivors have to pay all our medical bills with no aid from the government. He showed me his ID card which had entitled him to have all his medical bills paid by the United States government. His father was a Marine who had died in Vietnam, so he and his mother received these benefits, his mother for the rest of her life. I began to wonder why we didn't have similar benefits. Had I been fighting for the Japanese I wouldn't have expected to receive anything even if I was an American citizen; but I am an American citizen who had not been allowed to return to the United States; I was only a little boy.

Arai asserted that he had no choice but to remain in Japan during the war, and that he was one hundred percent American. As such, he strongly believed he had a right to compensation for injuries that resulted from an act of war conducted by his government.

The next witness was Tomoe Okai, president and founder of the Committee of Atomic Bomb Survivors in the U.S. She testified about her efforts to organize and obtain aid for the victims. She explained that in Japan atomic-bomb survivors were provided with medical treatment and other special considerations, while in the United States nothing of the sort was provided. Hibakusha were forced to pay high fees for medical examinations and treatment, even though American doctors had neither understanding nor experience in treating radiation illnesses. If a person was known to be a hibakusha, some life insurance companies would not accept him or her. She stated that the victims would achieve some peace of mind if only they could be provided with the same kind of medical treatment that is available to Japanese citizens. She concluded her statement by saying, "I sincerely hope that this kind of medical care is given to the atom bomb survivors in the U.S. so that they may be able to live in peace with this understanding and cooperation. This is my sincere hope."

The next hibakusha to testify was Kanji Kuramoto, chairman of the Northern California Committee for Atomic Bomb Survivors. Kuramoto, a Kibei Nisei, had been born in Hawaii. When the atomic bomb was dropped on Hiroshima, he was working in the mobilized student labor force in the city of Hikari. For two weeks after the bomb he searched among the ruins for his father, and it was during this time that he may have been exposed to residual radiation. Calling the dropping of the atomic bomb "the greatest crime ever committed by human beings," he related his grief-stricken and futile attempts to locate his father.

Kuramoto stated that he had been blessed with good health and a happy family after returning to the United States. At first, he said, he had tried to forget the tragedy of the bombing, but he was now extending his efforts to help others because he could understand the agony of the hibakusha, who were completely deserted by the American and Japanese communities. He found that pessimism prevailed among the hibakusha, of which there were fewer than one thousand in the United States, with under five hundred living in California. With such small numbers, they were politically powerless. The majority of the hibakusha were women who were neither assertive nor insistent and who, because they could barely speak English, were unable to convey their problems to their doctors or to the public. Because they were not wealthy, they did not have funds to promote their cause. For thirty years they had been ignored and forgotten.

"Please give them the aid they truly need," Kuramoto pleaded. "They cannot wait any longer. Please set up a program similar to the one instituted in Japan. I do not want to beg for your support, but I am appealing to open your hearts to aid these people in the spirit of true love."

Kaz Suyeishi took the stand next. She testified from her own experience that she couldn't buy insurance because she had been suffering from atomic-bomb disease. The hibakusha's "uneasiness" would be much relieved, she said, if a specialist in atomic-bomb illnesses could come from Hiroshima even once a year to give them medical examinations. She told how she had endured a heavy burden of anxiety ever since 1951, when she suffered an acute nervous breakdown while studying in Hawaii.

The last of the scheduled witnesses was George Morimoto, who explained, "I am speaking for my wife, Shigeko, who is too nervous to speak to you today." The statement he read told how she became a bomb victim as a twenty-one-year-old Japanese citizen. She lost the use of her right hand, and she had been discriminated against in trying to find work after coming to the United States. Her testimony revealed that the mental state of the hibakusha was one of powerlessness and disappointment. At the time, her husband's company health insurance covered her medical expenses, but when he retired, there would be no insurance, and there was always the danger that he would lose his job. In Japan, Shigeko was able to receive medical treatment at the Atomic Bomb Hospital, but she was living in America and it was almost impossible for her to go back to Japan for treatment. Moreover, in Japan there was the feeling that America had adequate doctors to care for the victims, so that it should not be necessary for the Japanese to provide treatment for American hibakusha.

> Many of us are fearful of coming forward because of family and social problems and the stigma already attached to this group. We are a relatively small group, but our needs are greater than the average. We need especially a medical plan that will assure each person in this group ongoing medical treatment and supervision. Psychiatric services will be a major need as families are confronted with more and more problems, both personal and as a group.

Questions that followed Morimoto's statement revealed that every segment of the population used public psychiatric clinics except the Japanese community. This exposed one previously hidden fact about Japanese people in America: They steer clear of psychiatric treatment, either because they feel no need for it, or because they feel ashamed of needing it.

The hearing was held as a forum for the hibakusha, and also for officials of concerned organizations and government offices who testified about the many problems the hibakusha were facing. Much of the testimony was reflected in a bill, later introduced in the state senate, to provide treatment and assistance to the hibakusha.

After the hearing, a working group was formed to draft a bill. The group was made up of members of CABSUS and representatives of the city and county of Los Angeles and the state of California. Noguchi served as liaison. Seven months later, on December 2, the bill was submitted as Senate Bill 15 (SB-15). Senator Dymally and Senator Song were co-sponsors, and Assemblyman Bill Greene of the state assembly was listed as an author. SB-15 called for the establishment of an institute for research and treatment for nuclear radiation, to treat victims of atomic radiation among the residents of California. These victims were to receive treatment at no cost. The chief stipulations of SB-15 were as follows:[2]

> The Legislature finds and declares that nuclear radiation, whether the result of accidents or warfare, inflicts tremendous disability on its victims. In California alone, there exist approximately five hundred citizens or permanent residents who suffer the physical and psychological effects of radiation. In addition, radiation threatens workers in California whose employment requires close proximity to nuclear reactors.
> It is, therefore, necessary, in order to reduce the severity of such physical and psychological effects and ease the human effect of radiation, that the

Legislature establish the California Institute for Research and Treatment for Nuclear Radiation.

The powers and duties of the institute shall be to:

1. Develop and establish a registry of atomic-bomb survivors and other persons affected by atomic radiation in California.
2. Issue a "certificate of nuclear radiation survivor" to identify persons who are eligible for treatment. The institute will determine whether the California resident is a nuclear radiation survivor with the burden of proof on the institute to show the applicant is not qualified.
3. Develop an information exchange program with scientists and experts on the treatment and research of atomic radiation.
4. Seek additional private and federal funds for the treatment and research of atomic radiation.
5. Employ persons fluent in foreign languages to communicate with California victims of atomic radiation who are bilingual.
6. Conduct seminars for health professionals specializing in atomic radiation.

Any California resident who suffers from atomic radiation as a result of exposure to atomic rays due to any wartime activity, or who was exposed to radiation on the job, or who was exposed to radiation by being in the vicinity of a nuclear radiation accident, or who is the natural child by birth of a parent who was in the vicinity of an atomic bombing or direct vicinity of a nuclear radiation accident, shall be eligible for treatment, at no cost, at the institute. Evidence of radiation exposure shall be substantiated by documents, records, witnesses or media coverage. If treatment cannot be obtained at the institute, the institute may contract with other appropriate agencies that provide the services of treatment for radiation exposure in order to assure proper medical care of such radiation victims.

Moreover, out of an advisory board consisting of nine members, the governor shall appoint one member who is a hibakusha, and another who is a social worker working with atomic-bomb survivors. And finally this bill appropriates $750,000 from the General Fund to the Board of Regents of the University of California for the purposes of this act.

This bill was the product of much painstaking work. Although it acknowledged the existence of hibakusha by referring to "five hundred citizens or permanent residents," neither "Hiroshima" nor "Nagasaki" appeared in the bill. The bomb itself was not mentioned and reference was made only to "wartime activity."

These omissions were intended to avoid the "Hiroshima complex," the tangle of guilt and justification that attaches to many Americans' response to the atomic bomb. The drafters of the bill thought it would be better to sidestep the question of the responsibility of the U.S. government, and to concentrate on obtaining treatment for the illnesses caused by the atomic bomb. Using the Japanese medical treatment law as a model, the bill prescribed that patients be registered and that certificates, or identification booklets, be issued. It was not required to have two witnesses verify a patient's eligibility, as required in Japan, because it would have been next to impossible to locate witnesses in the United States.

It is important to note that treatment was to be guaranteed also for the children of hibakusha. Hibakusha in the United States felt great distress because the genetic effects of radiation had not yet been determined, and it was not unusual for hibakusha to report that their children had a tendency to suffer from low blood pressure, frequent nosebleeds, and generally poor health.

The second strategy employed in the bill's language was to combine the plight of hibakusha with that of workers in nuclear industries who were also exposed to radiation. Nuclear power plants were already numerous in California, and the emergence of a class of nuclear accident victims could be anticipated. In the state senate public hearing, Dr. Franz Bauer, Dean of the School of Medicine at the University of Southern California, testified as a "staunch supporter of nuclear development which was recognized to be safe and efficient" and stressed the need for a statewide comprehensive safeguard system for workers in nuclear industry as well as for the general population. Bauer also testified to the importance of monitoring low-level radiation, in order to prevent cancer and other diseases in the future. It may have been questionable logic to combine past victims (hibakusha) and future victims, making them all beneficiaries of medical treatment at public expense, but it was thought to be a wise strategy to define victims of nuclear radiation as widely as possible in an effort not to focus solely on nuclear weapons.

Chairman Dymally wanted to locate the proposed institute at the University of California in Los Angeles. Towards the close of the public hearing he stated: "What I have in mind is a radiation clinic and treatment center administered by the University of California School of Medicine in Los Angeles and in San Francisco. The reason for this is because I think the legislature would respond favorably to the university system rather than any other operation."[3]

Prospects for the bill getting passed were slim. First of all, Senator Dymally, the chief sponsor of the bill, was elected lieutenant governor in the November 1974 election, and therefore became ineligible to deal with the bill. Under customary procedure, bills concerning minorities were taken care of by legislators representing minorities. This left control of SB-15 in the hands of Senator Song, but it was questionable whether his enthusiasm would match that of Senator Dymally.

Because Ronald Reagan was preparing for the presidential campaign in two years, he did not run for governor in 1974. His successor was thirty-seven-year-old Jerry Brown, whose father Edmund had preceded Reagan as governor. Brown's new administration was burdened by many problems, the biggest of which was an economic recession. His watchword was "tight budget" and under his administration there was no prospect of getting a bill passed that called for even modest expenditures. SB-15 would have required an appropriation of $750,000 from the general account.

At the beginning of 1975, SB-15 was sent to the Health and Welfare Committee of the senate. Senator Song, now the sole sponsor of the bill, took no action in spite of the wishes of the hibakusha. In a letter to the senator, Kuramoto, chairman of the northern California CABSUS, wrote:

> We, hibakusha, have waited for the last thirty years for legislation to provide medical treatment and to alleviate constant fears. We are prepared to lobby for the support of Governor Brown and the members of the legislature; however, we need your active support and advice as to what we should do.
>
> On behalf of the Committee of Atomic Bomb Survivors, I would like to urge you again to make the passing of the bill SB-15 your top priority in the Legislature. We need assistance from a humane politician like you.[4]

Kuramoto appealed to Song to hold a committee hearing as a first step toward passage of the bill. Song replied as follows:

> Dear Mr. Kuramoto:
>
> Thank you for your letter regarding SB-15 on Atomic Bomb Survivors. As you know, I became involved in this legislation with Lt. Governor Mervyn Dymally. In putting SB-15 together, neither Lt. Governor nor I had taken into consideration the condition of the State Budget or the wishes of the new Administration.
>
> Since we introduced the bill, Governor Brown has made it very clear that he will reject any proposal not covered in his budget. He is refusing, and rightly so, to expect state expenditure while our economic situation is in its present condition. For this reason the Lt. Governor and I agree that the bill has no chance this year. While we have not changed our minds on the bill's merits, we believe that it cannot become law until our economy markedly improves.[5]

Song thus distanced himself from the bill. "In the end," one atomic-bomb survivor lamented, "Senator Song did not do anything."[6] Some suspected that Song remained inactive on the bill because he had not enjoyed strong voter support from the Japanese American community in his district. But the story may have been more complicated than that. The above hibakusha continued: "He did not say anything to us openly, but after looking into it, we discovered he had said that since the Japanese government did nothing for Korean atomic-bomb survivors, why should he do anything for the Japanese in America?"

The Japanese government did not provide aid to the survivors in America, although there were still quite a few Japanese citizens among them, but it had also ignored Korean atomic-bomb survivors. In an ironic twist, this fact might now have hurt the cause of the American survivors. The hibakusha complained, "Atomic-bomb survivors in America are in an extremely unfavorable situation."

Unable to give up, the hibakusha began to look for a new sponsor for SB-15. In the meantime, Bill Greene, the state assemblyman who had co-authored the bill, had been elected to the state senate. Greene and San Francisco's Senator George Moscone became the new joint sponsors of the bill.

Just as hope for the enactment of SB-15 had begun to revive, a voice of opposition was raised from a totally unexpected quarter—the University of California. As one of the framers of the bill, Lieutenant Governor Dymally had believed that establishing a treatment center within the university system would help win the support of the legislature, but now the university itself raised objections.

Jay D. Michael, the vice president of the University of California in charge of government relations, sent a lengthy letter to Senator Greene, dated May 12, asking him to withdraw the bill or drastically revise it. He listed the following reasons for the university's opposition:

> In making this request, we are not unmindful of the plight of California residents suffering the effects of wartime exposure in Japan to nuclear bomb radiation to whose problems the bill is presumably directed. On the contrary, this plight is real and deserving of remedy. For the reasons set forth below, however, SB-15 is not the most effective solution to their problems and would, at substantial and unnecessary cost, duplicate much current treatment provision and research. . . .
>
> With respect to all victims other than those exposed to wartime radiation, medical care is already paid for by workmen's compensation or other private liability insurance which must be carried by the nuclear industry. Hence, to provide them with free care would add nothing to what they now have, and would be an unjustifiable windfall to the third party carriers who presently pay for their care.
>
> The only group not presently clearly provided for consists of individuals exposed to wartime radiation in Japan now living in this country. (The United States government provides care to such persons who remained in Japan and it is arguable that the federal government should extend this benefit to those who have come to the United States.)
>
> In any event, should the federal government refuse to do so, the interest of these individuals would be far better served by the state payment for their care wherever they choose to receive it, not at a single location only. There are several public and private facilities in California with expertise in this field, and the victims themselves are not concentrated in a single location.[7]

Like many Americans, Michael mistakenly believed that the U.S. government paid for medical care for hibakusha in Japan. His letter also argued that an increase in the research budget was not justified, because there was already a large number of research projects within the University of California system:

> The federally supported UC-San Francisco Radiobiology Laboratory has a current annual operating budget of $1.2 million; the UCLA Laboratory of Nuclear Medicine and Radiation Biology is currently conducting four programs of cellular research related to radiation exposure totaling $286,000 in the current year; the University's Lawrence-Livermore Laboratory has underway three programs related to nuclear radiation totaling some $250,000 this year; the Los Alamos Laboratory is conducting radiation-related biomedical and environmental research in some twenty-five projects funded in the current year at $5,058,000.

Michael also pointed out that in addition to the above facilities, expertise regarding accidental radiation exposure was being developed at the Lawrence and Donner laboratories at UC-Berkeley.

This strong opposition from the University of California forced proponents of the bill to abandon their original plan and revise the bill to focus on state support for the medical care of the hibakusha. The new draft of SB-15 proposed that Medi-Cal, the medical aid program in the state, be applied to hibakusha (including future hibakusha), by exempting the survivors from the low-income requirement for eligibility.[8] Medi-Cal was normally only available to the indigent and patients who were suffering incurable diseases. Most hibakusha were employed or supported by spouses who were working, so their income level left them ineligible for Medi-Cal benefits.

In general, Japanese Americans carried a strong bias against dependence on social welfare. Traditional Japanese ethics held that reliance upon government authorities is demeaning, and this value was held strongly by Issei and Nisei alike. Many of the Nisei hibakusha, in particular, had spent formative years in Japan, and their attitudes were very similar to those of the Issei. Japanese Americans were often held up as a model group among minorities in America, since they had relatively high incomes and a low percentage of people utilizing social welfare. The idea of free medical treatment was, in principle, difficult for many Nisei to countenance.

The primary reason the hibakusha accepted the notion of coverage under Medi-Cal was the great anxiety they felt about unpredictable future medical expenses. There was also the feeling that they were justified in receiving free medical care because their exposure to radiation from the A-bomb was a result of military action by their own government. They felt that someone, either the federal or the state government, had to take responsibility.

The key article of the revised SB-15 stated:

> Any California resident who suffers from atomic radiation as a result of exposure to atomic rays due to any wartime activity, or who was exposed to radiation on the job, or who was exposed to radiation by being in the vicinity of a nuclear radiation accident, or who is the natural child by birth of a parent who was in the vicinity of an atomic bombing or direct vicinity of a nuclear radiation accident, shall be eligible for treatment under Medi-Cal.

This bill was clearly a retreat from the original, because it did not guarantee access to a medical center with specialists in radiation diseases. In addition, the original bill would have provided full-time social workers who could understand the hibakusha's problems and who could speak Japanese.

However, this weaker bill was better than nothing at all. Even if the hibakusha were not completely satisfied, they could use it as a stepping stone in future organizing, reasoned those involved in redrafting the bill.

Kaz Suyeishi had gone to Sacramento in April to lobby in support of the bill. It was her first trip to the state capitol, and she was met at the airport by Paul Bannai, a Japanese American assemblyman. She spent the day making the rounds of concerned representatives.

"The fact that I spoke broken English only seemed to make them more sympathetic," she recalled. "The number of those who wanted to help increased. Some legislators were not simply sympathetic, but responded matter-of-factly that the bill should be brought to public hearing."[9]

A public hearing on SB-15 was held by the Senate Health and Welfare Committee on May 14th, and Suyeishi returned to Sacramento to testify. As the only hibakusha witness, she felt the burden of testifying on her shoulders, and she became dizzy from the tension. In front of the legislators and members of her organization, she described the thirty years of anxiety and distress the hibakusha had suffered. The legislators listened intently, and recommended passage of the bill by a margin of seven to one.

But this was only the first step. The members of the Health and Welfare Committee were predisposed to be supportive, and they had a good understanding of the bill. A second public hearing before the Finance Committee, which decided on budget appropriations, would be a more difficult hurdle. Still, the hibakusha were incredibly optimistic. One Japanese newspaper in San Francisco even declared, "The prospect for passage of the bill is bright."

On June 2, the Senate Finance Committee hearing was held at the state capitol with Anthony Bailenson as chairman.[10] Kuramoto testified, representing the hibakusha. The tone of his testimony was tense and sorrowful from the beginning.

> It was a terrible nightmare when I came back to Hiroshima two days after the atomic bomb was dropped. The devastated city was like a picture of hell. I cannot explain with my own words what I saw there. It was too much! It was really mankind's greatest sin. Near the epicenter I dug up over twenty dead bodies during my two week search for my lost father. I saw many dying victims on the ground. Most were suffering from burns.

Kuramoto was probably not observing the legislators' faces, but those who were watching their expressions would have noticed a response that was far from sympathetic. Their mood continued to harden.

> Two years ago, I met one of the members of the committee of A-bomb survivors in Los Angeles, and I found that there are still many untreated victims of the A-bomb, survivors who are suffering from incurable sicknesses. . . . Survivors of the A-bomb don't like to talk about their horrifying experiences, but one lady told me the terrible agony she went through.

Here Senator Bailenson interrupted Kuramoto's testimony, telling him that he had become too passionate and that emotional testimony would not be allowed.

The legislators began their deliberations. The chairman was of the opinion that the federal government should be responsible for compensation of the hibakusha, and that it was not the responsibility of the state government. The argument was advanced that if the state had to cover the medical expenses, the cost of the bill would be extremely high. If the bill were passed, it was suggested, hibakusha in Japan would rush to California to take advantage of the excellent medical care.

All of a sudden, someone shouted, "These people were our enemies!"

Kuramoto recalled that he began to shiver uncontrollably.[11] "What a statement!" he thought. "Were we the enemy? No, we couldn't have been. We were citizens born in America, and we only happened to be in Japan when the war broke out. How can they call us 'enemy' when we were injured by our own country's A-bomb?" But more than one person called the hibakusha "enemy" that day.

"They were our enemies. Why should we help these people?" one legislator opined, and another followed with a similar comment. In vain, a third legislator interjected, "I have a friend who is an American-born Nisei. Because of the war, he wasn't able to return to the U.S., but he's a U.S. citizen." But the tenor of the committee had already been set. When the vote was taken, SB-15 was defeated, seven to four. The public explanation was that the state's budget was not large enough, and that compensation should be carried out by the federal government.

In addition to the defeat of the bill, the biggest blow to Kuramoto and the other hibakusha was having the word "enemy" thrown at them. The word reflected prejudice against the Nikkei. Even though thirty years had passed since the war, the hibakusha were still regarded as enemies. That these people would call American-born citizens "enemies" just because they were of Japanese ancestry and happened to be in Japan during the war was nothing other than racial discrimination.

Two days later hibakusha, stunned by the double shock of the vote and the "enemy" epithet, visited the legislators who had voted against the bill, especially those who had called them "enemies." They were met with hard hearts and the response, "I will not change my opinion that you were our enemies. I do not see any need to change it."

Kaz Suyeishi painted a picture of bamboo and sparrows on Japanese rice paper and wrote a letter in both English and Japanese to the legislators, in which she wrote, "I am an American citizen born in Pasadena and proud of the United States. Please reconsider the bill on the basis of new understanding."[12] She got no reply. Her wish that she might become friends with those who had called her "enemy" went unfulfilled.

Kuramoto sent the following memorandum to the two Japanese American members of the State Assembly:

> We need your support in order to inform our leaders at Sacramento. We are not old "enemies" but good Americans. We know that this bill is very hard to pass. But we want to obtain at least a little conscience and kindness. We believe there are many humanitarians in the capital of our great State. I hope they understand that we, A-Bomb survivors in the U.S., are not "kojiki" (beggars). Thank you.[13]

After the bill was rejected, however, there was an increasing number of hate calls to Kuramoto's home, threatening him unless he stopped the effort. Most of the callers said things like, "Japs are Japs," "Remember Pearl Harbor!" or "Go home, Jap." They never identified themselves. They just hung up.[14]

Did Americans automatically think "Pearl Harbor" when talk came around to "Hiroshima"? As "forgotten Americans," the hibakusha had begun to move in search of their own identity. Now they were being called "the enemy." How could they "go home" when America was their native land? The hibakusha felt themselves falling farther into that deep crevice between the country of their parents and the country of their birth.

Notes

1. All the accounts are from the State Senate Hearings.
2. The text of SB-15 is in Kanji Kuramoto's personal papers (hereafter cited as "the Kuramoto files").

3. State Senate Hearings, p. 71.
4. Copies of his correspondence are in the Kuramoto files.
5. Kuramoto files.
6. Interview with an anonymous survivor, April 2, 1978.
7. Copy of Michael's letter to Green is in the Kuramoto files.
8. The text of revised SB-15 is in the Kuramoto files.
9. Interview with Suyeishi.
10. Partial record of the State Finance Committee is found in the Kuramoto files.
11. Kuramoto interview.
12. Suyeishi interview.
13. Copies of Kuramoto's memos are in the Kuramoto files.
14. Kuramoto interview.

twelve

In Search of Hibakusha

One day in April 1974, a request came to the Atomic Bomb Casualties Commission (ABCC) in Hiroshima from the Oak Ridge National Laboratory in Tennessee, asking that Hiroaki Yamada, survey section chief of the department of epidemiology and statistics, be dispatched to the United States.[1]

The town of Oak Ridge was created entirely from scratch by the Manhattan Project, in the midst of the Tennessee mountain wilderness. It consists almost entirely of facilities for the study and production of nuclear materials. The uranium used in the atomic bomb that was dropped on Hiroshima was produced here. The facilities came to be known, after the war, as the Oak Ridge National Laboratory, managed by the Union Carbide Company under a contract with the Atomic Energy Commission. The AEC had assigned the national laboratory the task of locating hibakusha throughout the United States. Yamada was asked to help with the task.

Yamada, a native of Hiroshima, had been in Korea at the time of the bombing. He had worked at the Hiroshima ABCC since its establishment in 1946. When John Auxier, chief of the Health Physics Division at Oak Ridge, conducted dose measurement tests with special reference to the factor of shielding in Hiroshima and Nagasaki, Yamada had accompanied and worked with him. Over several years of working in these cities, Yamada had become thoroughly familiar with all of the circumstances relating to the atomic bomb explosions, and he had acquired expert interviewing techniques. In addition, he and his counterpart from the Nagasaki ABCC had spent a year at Oak Ridge, beginning in late 1970. There he had studied methods for measuring radioactivity doses in living organisms. These tests, performed by the Health Physics Division, are among the most important functions of the Oak Ridge National Laboratory.

The measurement of radiation doses was—and is—no easy task. The dosage of radioactivity received by the hibakusha, for example, differed markedly, depending on the type and size of the nuclear explosion as well as the individual's circumstances: the length of the period of exposure, distance from the center of the explosion, and the nature of any shielding, such as walls, ceilings, and fences separating the individual from the blast. People in wooden Japanese houses received significantly more radiation than others, who were in the basements of concrete structures, even at the same distance from the hypocenter.

The reason it was considered important to measure radiation doses is that there was a scientific consensus that the risk to health varies in direct proportion to the amount of radiation exposure. Estimating dosage levels was part of an effort to determine "safe" levels of exposure, or the "maximum permissible dose." This endeavor was important beyond its meaning to those exposed to the atomic bomb explosions, because of the spread of nuclear power plants and the increased manufacture of nuclear weapons (along with various medical uses of radiation). The increase in the use of radioactive materials inevitably posed the danger of excessive exposure to larger numbers of people. The concept of maximum permissible dose is essential if human beings are to "coexist" with nuclear materials.

At the same time that Yamada was working on dose measurement at Oak Ridge, other scientists were disputing the basic premise that there were "permissible" doses of radiation. John Gofman and Arthur Tamplin, both eminent radiation biologists once associated with the Lawrence Livermore Laboratory in California, the Mecca of nuclear arms development, published a book that warned of the dangers of low-level radioactivity to the human body.[2] Another voice of caution came from Karl Morgan, former chief of the Health Physics Division at Oak Ridge and then professor at the Georgia Institute of Technology, who testified before Congress that there was no "safe" level of radioactivity, no matter how low.

It was natural that the ABCCs in Hiroshima and Nagasaki worked in cooperation with Oak Ridge. Among the categories of people who had been exposed to ionizing radiation during these years were the hibakusha, of whom 285,000 were identified as still living in 1950; American soldiers involved in later test detonations of nuclear bombs; the inhabitants of the Marshall Islands in the Pacific who were present during the nuclear bomb tests of the 1950s; and the increasing number of employees in the nuclear energy industry who were being exposed to radiation.

No data on hibakusha residing in the United States had been kept by either branch of the ABCC. One reason for this was that many hibakusha in the United States, including the large majority of Kibei survivors, left Japan before 1950, while the ABCC's population research figures were based on the 1950 Japanese national census. After that year, records of current addresses were kept for Japanese American hibakusha who returned to the United States and Japanese hibakusha who emigrated (mostly as a result of marriage).

From the viewpoint of the ABCC, the number of hibakusha who had returned to the United States and thus eluded their medical screening was considered insignificant. In any case, they were too widely dispersed throughout the country to be easily traced. No American medical or research organization had made a survey of the U.S. hibakusha, either. As far as the ABCC's scrutiny was concerned, they were "lost Americans." Some U.S. hibakusha had visited Japan and, while there, had been given physical examinations by the ABCC. However, none of them had met Hiroaki Yamada until his study sojourn in the United States.

Yamada was a logical choice as the person to conduct the survey of the American hibakusha because of his experience in conducting such surveys in Japan. Even before this appointment, Yamada had known of the existence of American hibakusha. At the end of his year of training at Oak Ridge, Auxier, the division chief, had asked him to make a point of meeting Thomas Noguchi in Los Angeles before he left for Japan. At about this time, in November 1971, the Committee of Atomic Bomb Survivors in the U.S. (CABSUS) had been formally established in Los Angeles. When Yamada came to L.A., Noguchi told him that he was determined to help the American hibakusha, and he later introduced him to Sumiko Tatematsu, the producer of the local Asahi Homecast radio program.

In their initial discussion, Noguchi was primarily interested in getting Yamada's advice on how to prepare a hibakusha survey questionnaire. Yamada emphasized the importance not only of determining the victims' distance from the hypocenter, but of how they were shielded from the explosion. The following day Yamada returned to Japan. The AEC was not yet aware of the existence of atomic bomb survivors in the United States, and Yamada did not initiate any project related to the American hibakusha. It took the formation of the CABSUS and its appeal through the national news media to bring the hibakusha to the attention of the AEC.

Scientists at the AEC began to explore the issue of the American hibakusha after August 1972, when they received the results of the mass medical examinations that the ABCC's Hiroshi Maki observed in Los Angeles. Noguchi and AEC personnel held their first exploratory talks in early October 1972.[3] In December, the house of delegates of the California Medical Association passed the resolution calling for assistance for hibakusha.

On February 27, 1973, the AEC decided to proceed with a scientific program of collecting radiation exposure information on the hibakusha in the Los Angeles area. This information was to be based on Maki's medical examinations at the Los Angeles County Hospital in August 1972. In April 1973, a preliminary study was completed by Oak Ridge in cooperation with the ABCCs in Japan. The study made a tentative estimate of the radiation dose received by each of forty-three survivors.

The following three steps were taken in tracking down the hibakusha in the United States: Preliminary information on survivors was collected by local volunteer groups; a search of the ABCC files was made for a dose estimate for each survivor; and a bilingual staff member of the ABCC interviewed those for whom there were no dose estimates. This last task was the principal responsibility of Yamada, who was not only proficient in English, but also in the Hiroshima dialect.

In January 1974, the AEC authorized an extension of the earlier Los Angeles study to other parts of the country. This was not difficult in the San Francisco area, where, in April 1974, the northern California branch of CABSUS had been organized, in part to aid and promote the survey. There were chapters in San Francisco, the East Bay area, San Jose, and Sacramento.

Japanese Americans in Hawaii, however, were surprisingly reluctant to participate in the survey. Their aversion to the whole hibakusha issue is worth a closer look.

Because of the large-scale emigration from Hiroshima Prefecture to Hawaii, beginning in 1885 and continuing for many years, there was little doubt that a large number of Hiroshima hibakusha would be living in Hawaii. The AEC estimated their number at about three hundred. When leaders of the local Japanese American community were approached about the survey, however, they suggested it would be fruitless, because few, if any, hibakusha were living in Hawaii. The survey, they intimated, might do more harm than good. Sidney Marks, who helped draft the survey plan and visited the University of Hawaii in Honolulu to talk with members of the local Japanese American community, recalled: "They were not enthusiastic about our project. Since we couldn't force their participation, we decided to drop Hawaii from the survey."[4]

Yamada remembered that the Oak Ridge staff told him that Hawaii's Japanese Americans had refused from the start to participate in the project. No one seemed to understand why they refused. Tōmin Harada, a Hiroshima surgeon and a well-known peace activist, who had a number of friends in Hawaii, offered an interesting observation: "People living in

Hawaii are warm-hearted and very loyal toward the U.S.," he said, "so they might have been inclined to dismiss the atomic bomb experience as trivial and not worth unearthing at this point."⁵

In general, hibakusha who succeeded in suppressing their experience of the bomb, or at least gave the appearance of having done so, were strong willed and came from relatively wealthy families. Perhaps the hibakusha in Hawaii were comfortable in their lives in the "Pacific Paradise," with its mild climate and tender-hearted people. They may have felt it was best to "let sleeping dogs lie." Also, many Hawaiians had experienced the Japanese attack on Pearl Harbor, and the U.S. military maintained a heavy presence and contributed substantially to the state's economy. Finally, one-third of the population in Hawaii is Japanese American. At that time, one of Hawaii's senators and both of its congressmen were Nisei. Japanese Americans were influential in state and local politics, business, finance, and education. A board member of the Japanese American Citizens League of San Francisco observed, "The greatest shock that Japanese Americans from Hawaii have in coming to the mainland is the sudden realization that here they are very much in the minority."⁶

On the mainland, the hibakusha's ordeal begins with the struggle to be recognized as an American by other Americans. In Hawaii, hibakusha do not have to struggle to be recognized as Americans. They are themselves members of a dominant minority. The Japanese American community in Hawaii might well have discouraged hibakusha from voicing complaints to the U.S. government, in fear of upsetting their well-established and nicely positioned apple cart.

The effort to locate hibakusha was thus restricted to the mainland. In September 1974, Hiroaki Yamada arrived in Oak Ridge to prepare for the survey. Although it was supposed to be nationwide, the survey would concentrate on areas with large numbers of Japanese Americans and, presumably, of hibakusha. The Committee of Atomic Bomb Survivors delivered a list of the names and addresses of confirmed hibakusha to George Kerr in the Health Physics Division at Oak Ridge. In an interview in the August 18, 1974, issue of the *National Enquirer*, William Doub, the AEC commissioner whose previous encounter with the supermarket tabloid has been noted, had declared, "So far we have located about 180 survivors and expect to have evaluated a total of 250 by year's end."⁷ The figure of 180 hibakusha was from California alone. The hardest part of the survey, yet to come, was to locate hibakusha outside of California, as well as hibakusha in California who were reluctant to come forward.

Yamada began his survey in Denver, Colorado, in November 1974, after his effort had been publicized in the local Japanese-language media.

Since the first decade of the twentieth century, a considerable Japanese community had grown up in Colorado, but the forced relocation required by Executive Order 9066 during World War II was not applied to the state. In fact, the Japanese population of Colorado increased substantially during the war. It was also likely that some hibakusha had settled in Colorado after marrying U.S. military personnel, since there had been a rapid expansion of military facilities there during the war and throughout the Cold War era. Yamada spent a day surveying Denver and another day in and around Salt Lake City, Utah, where there was also a relatively high concentration of Japanese Americans.

From Salt Lake City, Yamada crossed the Sierra Nevada mountains and arrived in San Francisco. Here Kanji Kuramoto and the Northern California Committee of Atomic Bomb Survivors extended assistance, and the city's public health division offered Yamada the use of an office and a telephone for making appointments and conducting interviews with hibakusha.

Yamada next traveled to Sacramento, the state capital, where there was also a concentration of Japanese Americans, particularly Hiroshima people. From there, Yamada passed through Fresno and headed south to Los Angeles, where he was assisted by Kaz Suyeishi and other members of CABSUS and by the Japanese Chamber of Commerce, which offered the use of an office. Southern California had the highest concentration of Japanese Americans on the mainland, but these citizens didn't, for the most part, live together in one community. They were scattered throughout the greater Los Angeles area, often at great distances from the city center, which made the survey very difficult.

The interviewing of hibakusha took Yamada much longer than he anticipated, because many of those he surveyed took the opportunity to unburden themselves of years of suppressed feelings. Normally, a scientific survey tries to obtain clear answers to a prepared questionnaire. But once the hibakusha began to speak with Yamada about things they had kept hidden for years, they could not easily be stopped. Interviews could take as long as an hour and a half. Yamada was primarily interested in getting people to talk about their circumstances at the time of the bombing, but many hibakusha had a desperate need to express their feelings.

A typical complaint: "Even my own son doesn't understand my feelings. I want people to listen and understand, but when I open my mouth, they give me a look that says, 'Not *gembaku* [the atomic bomb] again!' People around me give me cold looks; even if they listen, they don't understand."

Or a Hiroshima emigre would tell the interviewer, "What do you think, Yamada-san? If I went back to Japan I'd get a hibakusha ID card, and there would be specialists there on radiation disease. I'd get better treatment there. I'm thinking about going back."

Yamada tried to talk hibakusha out of this, advising them that it would make more sense to try to make the best of their present circumstances, since Japan had changed so much since they left. Looking back on the survey, he smiled wryly. "Sometimes I talked to them like a Buddhist monk. Many of them," he added, "definitely need psychological help." Frequently he played the role of psychiatrist.

Being fluent in the Hiroshima dialect was invaluable. A hibakusha told him, "Since I came back to the States, you are the first person who has really understood me." Quite a few hibakusha called Kanji Kuramoto to express their gratitude for Yamada's help and to ask for another chance to meet with him.

No matter how beneficial it was for the hibakusha's mental health, the outpouring of their long-suppressed feelings would never show up in the report of Yamada's survey, which was aimed solely at assessing radiation doses. Information about the hibakusha's state of health, their medical expenses, and health insurance problems was likewise left out of the report. The survey was defined only in terms of "health physics": determining location at the time of the explosion, in order to estimate radiation dosage. Yet AEC Commissioner Doub commented in the *National Enquirer* interview cited earlier: "It is not only the actual radiation dosage we are concerned with, but the mental torture that some of them might have suffered thinking they were exposed to radioactivity." But Sidney Marks, the AEC staff member who drafted the survey, contradicted this statement. "We were not allowed to determine any psychological effects of exposure to the atomic bomb explosion," he observed.[8]

It is probably impossible to assess the depth of the psychic wounds suffered by the hibakusha. There are no psychological equivalents of radiation dose units. Therefore, the scientists of the AEC, the Oak Ridge laboratory, and the ABCCs in Japan limited themselves to measuring the strictly quantifiable radiation dose received. As Marks put it: "If data can't be tabulated statistically, the truth can't be known. Laymen should receive explanations of such phenomena from experts."[9] When it came to the physical and psychological effects of the atomic bomb, however, were the hibakusha merely "laymen" without special knowledge?

Yamada returned to Oak Ridge just before Christmas 1974, leaving many hibakusha still to be interviewed. He confessed that he had lost nine pounds during his one-month trip. After several months, during which he recovered his strength and organized the data from the previous survey, he departed again for the West Coast in March 1975 to complete the bulk of the interviewing in San Francisco, Sacramento, San Jose, and Los

Angeles. This would entail another month's work. Three hibakusha lived in Eureka, California, near the Oregon state line, but, Yamada confessed, "I excused myself from traveling there and to some other areas by doing telephone interviews." Once again playing the dual role of Buddhist monk and psychiatrist, he often talked over the phone for more than an hour with each person. As Marks put it, "The survey was far more difficult than anticipated. We failed to foresee the magnitude of the problem."

The survey reached over three hundred hibakusha, well exceeding Doub's estimate. Working with the Committee of Atomic Bomb Survivors, Yamada had collected most of the data. He then returned to Oak Ridge to concentrate, along with George Kerr, on the complicated task of calculating the received radiation dose for each hibakusha. The survey report was completed in November, by the joint authorship of Kerr, Yamada, and Marks under the title, *A Survey of Radiation Doses Received by Atomic Bomb Survivors Residing in the United States*. It was the first documentation by a U.S. government agency of the existence of hibakusha in the United States. (By this time the AEC had been reorganized as the Energy Research and Development Administration [ERDA]. The Hiroshima and Nagasaki ABCCs had also been reorganized and now were jointly operated by the United States and Japan as the Radiation Effects Research Foundation [RERF].)

Kuramoto, Suyeishi, and the members of the Committee of Atomic Bomb Survivors looked forward eagerly to seeing the report made public. Earlier in 1975, the bill to provide aid to the hibakusha had been defeated in the California State Senate. As a result, the publication of the report was eagerly awaited as a morale booster by the membership of CABSUS.

In Japan, on February 1, 1976, the Hiroshima newspaper *Chūgoku Shimbun* published the entire report under the by-line of its New York correspondent, Akira Yamanaka. The article was prefaced with the question: "Though completed last November, this report has not yet been made public, even in hibakusha circles. Why?"[10]

By way of reply, the chief administrators of ERDA stated that they had been asked by supporters of hibakusha to avoid release of the report until after December 7, the anniversary of Pearl Harbor, out of consideration for the feelings of the U.S. public. They added that they also needed time to plan their follow-up to publication of the report.

Accounts vary regarding the real reasons for the delay, although it seems to be true that Thomas Noguchi did suggest avoiding the Pearl Harbor anniversary. Some members of the CABSUS regretted the time that was lost in making the report public. Many felt that it would have been proper to release the report at a press conference in Washington, but nothing of the sort took place. The report was not officially released until October 1976, and, except for the *Chūgoku Shimbun* article, no noticeable attention was paid either in Japan or in America.

What picture emerged from this survey of three hundred hibakusha? First of all, regarding the radiation dosages that were the primary interest of the survey, 151 of the survivors were in the "proximal" group, in other words, they had been within a distance of 2,500 meters (about 1.5 miles) of the hypocenter. Twelve of these had received 100 rads or more[11]; three had received 100–200 rads; nine had received 200 or more rads, including one who had received the abnormally high dose of 510 rads.

The overwhelming majority, however, had received less than 100 rads. Of these, five had received 50–100 rads; sixteen had received 20–50 rads; eighteen had received 10–20 rads; twenty-four had received 5–10 rads; forty-nine had received 1–5 rads; and twenty-six had received less than 1 rad. There was one case in which it was impossible to determine the level of received radiation.

The 120 survivors in the "distal" group, situated between 2,500 and 10,000 meters (1.5 to 6 miles) from the hypocenter, had received less than 1 rad. The other twenty-nine survivors were in the "not-in-city" group (having been over 10,000 meters from the hypocenter). These were classed as not having received any significant amount of radiation.

Along with the distance from the hypocenter, the presence and nature of any shielding were major factors in the calculation of the radiation dosages. For the sixty survivors who had experienced the radioactive "black rain" that occurred after the explosions, another 50 rads was added for Nagasaki survivors and 10 rads for Hiroshima survivors. It was also determined that neutron-related radioactivity had been significant only during a three-day period after the explosion and only within a distance of about 500 meters from the hypocenter. Twenty-seven survivors who were exposed in distant areas but entered soon afterwards into the area adjacent to the hypocenter were estimated to have received 5 rads or less of gamma-ray radiation.

Determining the received radiation dose must have been an enormously hard task because of the many factors involved in each case. However, the radiation received by the hibakusha who participated in the survey was surprisingly lower than they themselves had expected. If the degree of health risk increases in proportion to the dose of radiation, did it not follow that many of the hibakusha had been worried unnecessarily? If this was the case, no effort was made to allay this anxiety. The hibakusha were not individually informed of the calculations of their radiation doses.

Regarding geographical distribution of the survey sample: California, which had the overwhelming majority of hibakusha, broke down as follows: Los Angeles, 148; San Francisco, 76; Sacramento, 25; San Jose, 23;

Fresno, 7; Eureka, 3; for a total of 282. The other states: Utah, 5; Colorado, 4; Oregon, 2; and Arkansas, Florida, Illinois, Michigan, Pennsylvania, Texas, and Washington, 1 each.

There was only one Caucasian hibakusha, a man in Texas who was an American prisoner of war in Nagasaki at the time of the bombing. As of 1976 he was still hospitalized in the Houston Army Hospital. The survivor's wife had called Oak Ridge to request an interview for her husband. According to Yamada, Kerr conducted the interview; further details are unknown. The subject's mental status was said to be unclear. From previous accounts of his story, it appears that he was doing forced labor in a coal mine, and the bomb exploded just as he emerged from the mine. He maintained that he was carried out by a freight train four days later and returned to the United States.

All the rest (299) of those included in the survey were of Japanese ancestry. Of these, 130, or forty-four percent, had been born in the United States, forty-nine male and eighty-one female. Forty-eight were naturalized American citizens of Japanese birth, eleven males and thirty-seven females, the ratio attributable to the frequency of marriage of Japanese women to U.S. citizens. The total number of American citizens within the entire hibakusha group was 178, or about sixty percent. The remainder of the group (except six males and five females who were unclassified) was composed of Japanese citizens with permanent resident status in the United States. These numbered 110, eighteen men and ninety-two women. Those married to U.S. citizens featured prominently in this group also.

Toward the end of the survey report, it was stated that this group of three hundred hibakusha cannot be considered a random sample of the total hibakusha population in the United States, because the survey was conducted on the basis of voluntary, not compulsory, participation. The report read, "Systematic search and persuasion was not employed out of regard for the individual sensitivities of the survivors."

Notes

1. This chapter is based on interviews with Hiroaki Yamada, March 4, 1977, and May 22, 1978, in Hiroshima; also George D. Kerr, Hiroaki Yamada, and Sydney Marks, "A Survey of Radiation Doses Received by Atomic Bomb Survivors Residing in the United States," *Health Physics* 31 (October 1976): 305–313.
2. John W. Gofman and Arthur R. Tamplin, *Poisoned Power* (Emmaus, PA: Podale Press, 1971).
3. Thomas Noguchi interview.
4. Telephone interview with Sidney Marks.
5. Interview with Tōmin Harada, May 21, 1978, Hiroshima.
6. Conversation with a JACL official, April 6, 1978, San Francisco.

7. *National Enquirer*, August 18, 1974.
8. Marks interview.
9. Marks interview.
10. *Chūgoku shimbun*, February 1, 1976.
11. The rad is the standard unit of absorbed dose of radiation, equal to the energy absorption of 100 ergs per gram (0.01 jougle per kilogram). According to studies of the ABCCs in Japan, there is an abnormally high occurrence of malignant lymphomas and bone marrow cancer in the group that received 100 rads or more, and a high incidence of uterine and breast cancers and leukemia in the group that received 200 rads or more.

thirteen

The Many Shades of the Hibakusha Experience

In the report based on Hiroaki Yamada's survey, hibakusha were reduced to abstract statistics, faceless human beings who carried within their bodies a certain amount of radioactivity. But, of course, each hibakusha did have a face, a body, a mind, and a unique way of responding to the experience of the bomb. Thus, the jigsaw puzzle comprising the one thousand hibakusha across the United States was not monochromatic. It contained many colors and shades.

Many of the hibakusha introduced up to this point have been rather grave and dark figures, but it is important to recognize that there were other survivors whose lives were not as overcast with gloom. Looking at some who were living relatively untroubled lives may help us better understand those who were tormented by the bomb. Most of the accounts that follow take up the stories of hibakusha who were introduced in the first chapter of this book. What kind of lives had they lived in the ensuing decades, and what were their feelings about the bomb?

Akira Furuta, it may be recalled, was born in Portland, Oregon, and was attending Hiroshima Second Middle School when the bomb was dropped.¹ Some time after the war he learned that he was the adopted son of an Issei who had been living in Tacoma, Washington. He discovered this fact when, discouraged about a future in Japan, he obtained his birth certificate, with the intention of going to the United States.

"Because my mother had died," he recalled,

I had no hope of going to college after I finished high school, so I went to work for a construction firm run by some relatives. But about that time my oldest sister, who was married to a Nisei, gave me some money so that I could go to college. I went to Tokyo in 1952 and started attending night classes at Meiji University, while working part time during the day.

I was working at a U.S. Army supply depot in Shinagawa, and some American officers asked me why I had not gone back to the States since I was born there. They told me that I would be drafted within six months if I went back, but then I could qualify for the GI Bill if I volunteered for service in Japan. I started the process in my sophomore year, only to find that my family register did not correspond with my U.S. birth certificate. Although I felt badly for my stepfather, I took him to court to prove that I was not his real son. My relatives were very angry. But there wasn't much of a future for me in Japan, even if I could graduate from college.

As a result of the court hearing, I lost my Japanese citizenship and temporarily became stateless. I applied for the restoration of my U.S. citizenship at the consulate in Yokohama, where a Japanese official scared me by telling me I would be fined for not having registered for military service before I was eighteen. In the end, I joined the army at Camp Drake in Asaka as a private in November 1954, and within two days I was flown to the U.S. in a military aircraft. There were no immigration formalities. I was sent to Ford Ord, where I received training.

It was really hard because I didn't understand English well. When they shouted 'Right face!,' I had to translate it into Japanese in my mind before I understood, so I was always one beat behind the others. My brother was against my joining the army, but I went ahead because I wanted the GI Bill and not because I felt any loyalty to the States. I only wanted to go to school in the U.S., and I intended to go back to Japan after graduation.

Furuta remained in the army four years, spending some time back in Japan at the base in Asaka, near Tokyo. While there he resumed his study at the Meiji University night school and graduated in 1956. He was stationed in Korea for some time, then discharged as a master sergeant in 1958. He attended the University of Washington on the GI Bill, earning a degree in mechanical engineering in three years (since this was the limit of his GI Bill benefits, he also attended school in the summer).

By that time Furuta had decided to settle in the United States, because of the good opportunities available to college graduates, and he went to work for an aircraft manufacturer in Seattle. "I had almost forgotten my Japanese," he noted, "but recently it has been coming back to me since I started doing things for the Japanese community."

"I was conscious of the fact that I was exposed to the atomic bomb," Furuta recalled,

> but I am very healthy and I have had none of the symptoms that many hibakusha suffer. The only experience I had as a hibakusha was in the service,

when I told a military doctor that I had been exposed to the atom bomb, and I was told that I didn't have to donate blood. I tell my doctors I am a hibakusha, and they give me a strange look. I don't like gloomy things by nature, you see, so I don't talk about it much.

While I was in service in Korea, I went back to Hiroshima on vacation. I had heard that people in Hiroshima were extremely critical of America because of the atomic bomb, and I was wearing my military uniform when I went there. An acquaintance, who knew I was a victim of the atomic bomb, told me he thought the United States was very wrong. He had been a master sergeant in the Japanese Imperial Army and was not in Hiroshima at the time of the atomic bomb. I responded that the Japanese army did terrible things in Nanking, too. 'You weren't here,' I told him, 'but I was exposed to the bomb. But you can't say just because of that I should not be wearing an American uniform.' I was a bit angry. When I think back on it now, it was a natural reaction for the people of Hiroshima. Japanese always have to blame someone for everything.

As to what meaning the hibakusha experience has had for me, the most important issue to consider is what actually led to the bombing. I recently heard that Churchill and Truman decided to use the atomic bombs in order to check the USSR, because Stalin was demanding Czechoslovakia and other countries in Europe. So the bomb was actually to stop the Soviet Union! There are two ways to look at this from a strategic point of view. People generally think of Hiroshima as a peaceful city, but actually it was a militarized city. The whole population was mobilized into the military. That's where America dropped the bomb. With normal weapons, guns and all, it's 'Bang, you're dead.' There are no lingering problems like radioactivity, which you have with an atomic bomb. That's what I hate about the atomic bomb.

But if I look at this from the American side, it appears entirely different. I think Truman had a point. If we had fought a decisive battle on the mainland like we did in Okinawa, I'm sure many people would have died. I don't buy the argument that Japan would have surrendered without the bomb because the Japanese were already losing. The Japanese were saying that they would fight to the last soldier. Until August 14, I believed Japan would never lose. If there had been a final battle on the mainland I was ready to do as I was told and fight to the end. Everyone I knew was ready, too.

I do not feel any irony of fate in what I have been through. Unlike the Nisei who were born and raised here [in the United States], I do not have a strong feeling of citizenship. Do I feel loyal to the United States? I don't know. American politics is opportunistic itself. If we ever have another war, I don't want to be on the losing side. Losing a war is a miserable experience, you know.

For the most part, the American hibakusha were situated in a chasm between Japan and the United States. But Akira Furuta had not fallen into that chasm. He seemed to have one foot firmly planted on either side and was living a full life. He had not suffered any aftereffects of the bomb and was both mentally and physically healthy. Furuta described himself as "having none of the typical conditions of a hibakusha," but his story is another piece in the picture of hibakusha in the United States.

～ ～ ～

Judy Enseki (who had gone from the Manzanar camp to Japan on the second exchange ship with her husband and baby boy) left Hiroshima in November 1945 for Tokyo, where she found work with the occupation forces.[2] Several times she applied for permission to return to the United States, but she had no means of proving her U.S. citizenship, and her applications were denied. However, many Nisei friends that she had known back home in Delano, California, had joined the U.S. forces and were in Tokyo with the occupation, so she no longer felt the loneliness of living in a "foreign land" that she had experienced during the war.

"But I knew that I had to get back to the United States," she recalled,

> because it's my country, you know. Then out of the blue the permission was granted, and I returned on an American President Line ship with a ticket my parents sent me. That was in April 1947. My husband was still detained in Siberia, so before I left I went to Hiroshima to say good-bye to my mother-in-law and to leave a letter for my husband. He came back to Japan in the fall of 1947. He remarried, and we divorced officially in 1950. I raised my son on my own, and put him through college, too. I think my husband never had any intention of coming back to the U.S.
>
> Outside the courtroom after I got my divorce, I was surrounded by reporters. My church had contacted them. There was a headline in the newspaper the next day: "Atom Bomb Witness Divorces Mate." I was so embarrassed. After that, I was often asked to appear on television on the anniversary of the bomb.
>
> How does that make me feel? I don't feel anything in particular. I don't have strong feelings about the atomic bomb or nuclear power. You know, there's no point in focusing on just one kind of weapon, it won't get us anywhere. We have to become more civilized, to the point where we outlaw man's inhumanity to man. That's got to be the starting point. Until we learn to be kind to one another, it won't do any good to ban nuclear weapons or destroy chemical weapons. Deciding that one weapon is OK and another is unacceptable is really missing the point altogether. Humans have to be less greedy and learn to understand each other's problems.
>
> I am not bitter at all about the bomb. I don't know if that is because of my Japanese upbringing or the influence of my parents, who were very optimistic people. You might say it's because neither my son nor I were injured by the bomb, and that might be true. I have never been conscious of being a hibakusha. It was only when I was asked to join the hibakusha group that I became aware of the hibakusha, and I offered to help. They mostly speak Japanese, so I don't understand their discussions, but when Kaz [Suyeishi] needs to write letters to senators, I type them up for her, because I used to be a secretary.

The Many Shades of the Hibakusha Experience

> What do I think about the past? I always felt I was in the wrong place. But now I'm in the right place. I'm not a person who dwells on the past. The future is what's important and interesting. If the past doesn't let you go, it's because you let it drag you down. I have too much fun thinking about what I am going to do tomorrow, next year, and the year after that.

At the time she was interviewed in summer 1976, Enseki was the director of a suicide prevention center in Los Angeles, living her life fully as an American citizen without any mental or physical effects of the atomic bomb. In terms of gradations of brightness in the jigsaw puzzle of the hibakusha in America, Judy Enseki would definitely be placed at the brightest end of the spectrum.

It is possible to argue that Furuta and Enseki are optimistic because they were suffering no aftereffects, even though they had been exposed to the bomb. Still, they had witnessed a living hell, and the brightness of their outlook could not be merely naive optimism. For instance, Furuta remarked, "I saw many people who had been hit by the bomb walking toward me with their skin hanging from their limbs. Even today when I see someone who has died, I don't feel much of a shock. I think that's because of the psychic blow from Hiroshima." This was not the observation of a Pollyanna.

How about Jane (Ishigame) Iwashika?[3] Born in Fresno, California, she not only witnessed the inferno in Hiroshima, but was also badly injured by the blast and hovered between life and death because of radiation disease. After the war, everyone in her family returned to Japan from the relocation camps in America. But Iwashika came back to the United States alone in 1948.

"From the beginning it was my intention to come back to the U.S.," she recalled.

> I had no emotional attachment to Japan. All my friends were dead. My older sister (Irene) had already come back to the States, and my grandfather told me he had some money over here that I could use go to school. I did go to school, but it didn't work out. I had to work to earn a living, and then I got married, so I never returned to school. I came back to Los Angeles in 1948 and married my husband in 1950. My health was bad; I was always suffering from one thing or another, morning and night. This continued until about 1960.
>
> I gave birth to my first child in 1951. I was still feeling sick. I did not think about the atomic bomb at all then. Even when I was still in Japan, nobody talked about the bomb. "Atomic bomb" was a "no-no" word. I just suffered by myself. But I got better as I got older. My second child was born in 1953. I developed strong will power, and I'd go to work even if I had to crawl there.

If my husband had been too nice to me, I probably would have stayed in bed. I had heart trouble that continued until several years ago. Even before my marriage, I went to dermatologists because of skin rashes. I had stomach problems for many years, and I can't walk straight. In 1957 I was diagnosed with tuberculosis and hospitalized for seven months. I showed no sign of improvement, and my doctor even asked me why I bothered to come to the hospital. My teeth were pulled in 1960 and I have false teeth now. My shoulder was dislocated by the bomb blast, and I put it back myself incorrectly, so I had to have an operation later to have a pin inserted to connect the bone and the muscles. I have myopia and presbyopia, and trouble with my nose, too. I wonder what my husband saw in me to marry!

But my medical care has been free, because I have been working at a supermarket. The Retail Clerks Union is a large and powerful organization, and I was covered by their group medical insurance. However, the insurance does not cover you unless you work full time, forty hours a week. I have to stand all day long, and although we get an hour for lunch, we actually only spend about ten minutes eating, the market is so insanely busy. But I am paid the same salary as a man. I often collapsed at work, but I never took the day off. All I'd ask for was a ten-minute break. It's really a matter of will. I don't care when I die, but I'll probably live to be a hundred. When I'm sick, I cry and cry, but then I stop and ask myself, "Why am I crying?"

I imagine the people in the hibakusha group are really sick, aren't they? I don't need sympathy, and I don't need a medical aid bill. Of course some people need it, but not everyone. Aren't they mostly Japanese citizens? I was hurt, too, but I'm still working hard so I can keep my insurance coverage. I want to be covered by insurance even after I retire, so I work even harder. I've already worked twenty years, but I can't get a pension yet because I am too young.

Those people pity themselves. My grandmother told me that if others laughed at my scarred face, I should laugh with them. If you feel happy, your face will look happy, too. I think what those people really need is psychological care rather than medical aid. The problems of the hibakusha are all in their heads. I felt that way when I took care of them some time ago. They're piling burdens on their own shoulders. Aren't they still fighting World War II? You should forget what happened yesterday. What's the use of looking back?

Back in Japan, my face was covered with scars and I looked awful. I often thought of committing suicide when I saw girls my age who looked beautiful. I was eighteen or nineteen then, my friends were all dead and I was alone. I had my parents but only in name. But these are things of the past. Now I live, finding joy in my own children. Of course I struggle, but if I feel sick, I just go to bed. You've got to rely on your own strength. If someone tells you you're sick, that's when you feel sick. I don't believe in any religion, but I have faith in myself.

What does the atomic bomb experience mean to me? It changed my whole life, I think. For better or worse, I'm not sure. I think it was for the better in the sense that it made me able to fight for myself. If I had not been a hibakusha, I would've become an ordinary housewife or a farmer in Japan. But because of the bomb, when I feel sick, I get my children to do lots of things, so our children are more mature than in normal families. I always try to think in a positive manner. When I get sick, I scold myself and call myself a weakling.

What do I think of the atomic bomb? That was war. You have to think that way. It is the will of God that a person has to die. Christ had to die, too, didn't he? Dying is not frightening at all.

Jane Iwashika exhibited a tranquillity of mind, like the calm after a storm. The suffering she had been through to drag herself from the devastation of Hiroshima to that state of mind is unimaginable. With her strong will and the support of her husband and children, she was able to pull through those decades. The problems of the hibakusha are most apparent in those people who do not have this type of support and inner strength to help them carry on.

The following is the story of another Nisei hibakusha woman who returned to the United States.[4] Yukiko Johnson (not her real name) has assimilated and adapted so thoroughly to American society that one hesitates to use the term "Kibei Nisei" to describe her. Among the many hibakusha in the United States, this survivor, with her weak hibakusha consciousness, seems to belong somewhere around the edge of the puzzle.

Johnson refused several times to be interviewed, on the grounds that she didn't want to have "anything to do with the hibakusha group." She finally agreed to see me, accompanied by her husband, for about one hour. To describe her in a most general way, Johnson was born in northern California, and she and her family moved to Japan right before the war broke out. She was a post–middle school student at the time that she and her mother were exposed to the atomic bomb at their home. The only injuries Johnson suffered were caused by flying glass. However, her grandmother was caught under a fallen crossbeam, and they were unable to free her. When the spreading fire approached them, she and her injured mother had no choice but to escape, leaving the grandmother behind. After the war, she graduated from college and worked for a U.S. government agency. Later she came to study in the United States and married a white technician she had met at work. They were living in an affluent California city. The following is taken from our conversation. (It should be noted that the interview was conducted in English. Her command of the language was exceptional.)

> I don't understand why people are so concerned with the atomic bomb experience. We couldn't help my grandmother who was burned to death after the house fell on top of her, but a lot of those things happened at the time of the

atomic bombing; it was just one of those things. Besides, my mother and I did our best to help grandmother out, but we just couldn't do anything. My father taught me to think about what I should do next when something happens. So I don't look back. Those things happened thirty-five years ago, and to me an automobile crash yesterday would be a lot more shocking. I didn't suffer any psychic trauma because of my experience of the bombing. My mind is never excited by it like other survivors. I sometimes recall it, but that's that.

Her husband noted, "She is the most unneurotic person I have ever met." Johnson continued: "I managed fine when I went to Japan from the States, too. My adaptation to the society was very smooth, and there was no problem at all. I don't remember going through a transition period from being a Japanese to an American after the war, either. I was always a U.S. citizen, you see."

This case can serve as one index of the hibakusha experience in the United States. If we take this survivor's words at face value, they imply that some people are capable of suppressing from their consciousness even such a terrifying experience as the atomic bomb, and that the bomb experience loses its meaning as one is assimilated into American society. This hibakusha had no difficulty in assimilating, and her hibakusha experience apparently did not cause any conflict in her mind.

At the other end of the spectrum is the following, a voice filled with bitterness toward the atomic bomb, welling up from deep inside the survivor. This is an unusual piece of testimony, because when their anger is so strong and deep, hibakusha are generally not inclined to talk about it. Sadako (Obata) Shimazaki was interviewed during her thirty-first summer in America.[5] She had never gone back to Japan since she returned to the United States, having given up her daughter, an in utero hibakusha, for adoption.

> I came back alone in 1947. After my parents got out of the camp in Poston, my father became a gardener in San Mateo, south of San Francisco, so I joined them there. Starting about that time, my blood pressure was very low and I was often dizzy. After graduating from sewing school in Los Angeles, I married a Nisei from this area. His former wife had died after she gave birth to a child, so we had similar situations. It was some twenty-five years ago that we came to Monterey Park. My husband was running a grocery shop here, and we hoped to build a home. But we were the first Japanese in the area, and racial discrimination was quite strong. We often received intimidating letters in our mailbox.
>
> Our first child was a daughter. She was healthy at first, but her health began to deteriorate when she was about fifteen, and she started to have fainting spells. Two years ago she was going to get married, but all of a sudden she could not stand up on her own feet. The cause was unknown. After

about three weeks in the hospital, she was able to get around a bit and she was discharged. She got married later that year. But the next year she developed an extremely high fever and got so sick that it interfered with her breathing. Again the cause of this illness was unknown, but I have a feeling that maybe it came from handling germs in the university laboratory.

Our next child was a boy, who had poor health from birth. His joints broke very easily because of hemophilia. He often stayed out of school, but he did well and got A's in all his subjects. He's very smart and, in fact, graduated from UCLA on a scholarship. He is in the hospital right now.

I had two operations for breast cancer, and now the cancer has spread to my bones. It gives me great pain, but I am enduring it because I'm told that I'm not dying right away. I couldn't sleep at night because of the pain, so I went to the doctor, and he told me there was cancer in the bones throughout my body. My doctor is watching me carefully, so that it won't spread further. He tells me that I don't have to worry too much about cancer that spreads from the breast to the bones. There is a spot on my lung, but we do not know if it is also cancer, or it might be one of the pieces of glass that were buried in my body during the atomic bombing, since these still come out of my body now and then.

When I ask my doctor if my son's and my own poor health have anything to do with the atomic bomb, he says, no, they don't. When Dr. Maki came from Japan, he examined me, but at that time a white doctor again told me that my sickness had nothing to do with the bomb. They don't even say "maybe." My son's hemophilia is a rare disease. Nobody else in my family ever had it. It is a disease that was common among royal families who interbreed, and I hear it is called the royal disease. We make jokes in our family, and tell ourselves that we are royalty.

My aunt, who was also a victim of the atomic bomb, died of cancer that spread all over her body, so I suppose I may end in the same way. Nobody in my family ever had cancer. I wonder if this really has nothing to do with the atomic bomb. Fortunately, my medical expenses are covered by insurance from the Kaiser Foundation, through my husband's work. He jokes that he might be kicked out of the program, but so far he hasn't. My son isn't covered, though, so we have to pay hospital costs. When we get someone to donate blood, it's OK, but otherwise we have to buy blood. The operations have cost us thirty-four thousand dollars so far. Japanese Americans in southern California have given us a lot of support. This time, my son was supposed to be in the hospital for eighteen days, but as of today it's been four weeks and I'm worried. My son's disease is a "crippling disease," which Medi-Cal covers, but the limit is fourteen days and after that it's up to the doctor to approve an extension.

We have two younger sons and our youngest child is a daughter. All of them have been fairly healthy. While I was giving birth to the children I was fine, but afterwards I commuted to the doctor for treatment of a stomach ulcer and gallstones. Doctors tell me I suffer from anxiety. Well, they're probably right. But I think the atomic bomb is the cause of my poor health. Why do these things happen only to me? I try not to be bitter. What's done is done. If I had not gone to Japan, this would never have happened to me, but I don't blame anyone. I've been to the hospital many times, and when I found out I had cancer for the second time, I cried like mad, thinking, "This is all because of the atomic bomb."

My story is endless. It's a living hell. I'm not working with the hibakusha group, because somehow I don't feel like going to their meetings, but I would like to have regular checkups. We can't just go back and forth to Japan every time we want treatment. I wish doctors would come from Japan. I really want to see a specialist. Doctors here don't really know anything. I don't have to worry about the cost of treatment thanks to the insurance, but I still want to see a Japanese doctor. Generally speaking, Americans don't like to talk about the atomic bomb. It's a touchy subject, you know. They are the ones who dropped it. But I wish they would at least build a special hospital.

Since I've never gone back to Japan, I don't have a hibakusha certificate. I am concerned about my daughter, the one I gave up for adoption, and I was planning to go to Japan in October. But I am not sure if I can, since the pain in my bones has lasted for three weeks now. My daughter in Japan was an in utero hibakusha, so she is a little retarded. She only finished eighth grade and learned sewing, then married. But her husband died, and I hear that she is having a hard time making a living. She had an operation for a brain tumor a while ago, and now she is staying home doing nothing much. I want to go to Japan and prove that she really was an in utero hibakusha, because she has written that it is hard for her to pay her medical expenses.

The more I think about atomic bombs, the more scared I get. I wish there had never been an atomic bomb. Any kind of bomb is terrible, but with these bombs it's the aftereffects that are frightening. There is no end to it. The future is so uncertain, I get scared thinking about what will happen. After the war, I tried hard to forget it and not to be bitter, but I am reminded every time I get sick. Questions about the bomb come to the surface—"Why Hiroshima? Why me?"

Notes

1. Interview with Akira Furuta.
2. Interview with Judy Enseki.
3. Interview with Jane (Ishigame) Iwashika.
4. Interview with Yukiko Johnson (pseudonym), September 21, 1977, San Francisco.
5. Interview with Sadako (Obata) Shimazaki.

fourteen

Ups and Downs

About the time Hiroaki Yamada from the Hiroshima ABCC returned to Oak Ridge after his second survey of mainland U.S. hibakusha, a diminutive woman was visiting survivors in California. Her name was Barbara Reynolds, or "Barbara-san," as she was known to everyone in Japan who was working to help the hibakusha. Unselfish, dedicated, and at times quite extraordinary in her actions, her work provides an example of how much a single determined individual can accomplish.[1]

Although both were American exports to Hiroshima, the activities of Barbara Reynolds and those of the ABCC were a study in contrasts. Reynolds was motivated by God's love for humanity; the ABCC investigations were conducted from the detached perspective of science. Yet it is interesting to note that Barbara Reynolds was introduced to Hiroshima through the office of the ABCC.

Barbara came to Hiroshima in 1951 with her anthropologist husband, Earle Reynolds, who had been invited by the National Academy of Sciences to do research in Hiroshima with the ABCC. During the three years they lived in American dependent housing eighteen miles from Hiroshima, Barbara was just another good-hearted American.

In the fall of 1954, the Reynoldses and two children sailed from Hiroshima on board a yacht, the *Phoenix of Hiroshima*, which had been built for a round-the-world return trip to the United States. When they completed their circumnavigation in Hawaii in June 1958, the Reynoldses heard that the United States was conducting a series of nuclear tests at Eniwetok Atoll in the South Pacific. Quakers aboard a vessel called the *Golden Rule* had announced their intention of sailing into the restricted waters in protest, but before they could set sail, their yacht was taken into custody. The entire crew of the *Golden Rule* was arrested and sentenced to sixty days in jail. Impressed by the dedication of the Quakers and following an inner voice that told them to "act," the Reynoldses decided to carry out the protest. In doing so, they deepened their understanding of the meaning of Hiroshima.

Earle Reynolds was arrested, tried, convicted, and sentenced to eighteen months in prison and five years of probation. The conviction, however, was overturned on appeal, delivering a blow to the tremendous power of the AEC. In 1960, the Reynoldses returned to Hiroshima as tried and tested advocates of peace.

Barbara Reynolds's activities after their return form part of the history of Hiroshima. Among the most memorable of these are two "peace pilgrimages" she led in 1962 and 1964. Both missions traveled around the world to all the nuclear nations—the first, with two survivors; and the second, with a larger group of people that included twenty-six hibakusha and twelve companion interpreters. Both of these missions brought the voice of Hiroshima to diverse audiences. As a legacy of the 1964 "World Peace Study Mission," the World Friendship Center was established in Hiroshima, with the help of Dr. Tōmin Harada and the mission participants.

The Reynoldses were eventually divorced, and Barbara continued to carry on her unique activities at the center. Every year, when August 6 came around and the fractious Japanese peace movement battled for prominence, Barbara quietly received guests from many countries and tried to help them understand Hiroshima. At Christmas, she would fast in the Peace Park, continuing her prayer for peace. She was the conscience of America in Hiroshima.

Reynolds left Hiroshima in the spring of 1969. The peace movement, and she herself, had reached a turning point. Harada, who had been chairman of the WFC's board of directors since its inception, assumed responsibility for guiding the center, and Barbara went to Pendle Hill, a Quaker center near Philadelphia, to study writing under the guidance of Elizabeth Gray Vining, who had been a tutor to the crown prince of Japan during the occupation. But Hiroshima, the subject she was determined to write about, was not popular in the United States, and she could not find a publisher interested in a book about the atomic bomb and its survivors.

For several years she attempted to find an audience for the voice of Hiroshima, while the attention of the American peace movement was focused on the Vietnam War. She was finally able to find a Quaker college that was interested in preserving her materials about Hiroshima and Nagasaki, and in 1973 she began work at the Peace Resource Center at Wilmington College in Ohio. There she established a memorial collection of all available information in both English and Japanese related to the damage caused by the atomic bombs, the ordeals of the hibakusha, and the anti-nuclear movement.

Ups and Downs   145

It was in the course of gathering material for the center that Reynolds learned about the U.S. hibakusha. Included among the miscellaneous materials sent to her in 1974, there was an article from the Los Angeles newspaper, *Rafu Shimpō* about the American atomic-bomb survivors. Reynolds immediately wrote to Kanji Kuramoto in San Francisco and asked for details about CABSUS. At that time, she was planning a peace education conference for the following August, the thirtieth anniversary of the bombing of Hiroshima and Nagasaki. Dr. Harada and many hibakusha from Japan were expected to attend the seminar, but if there were atomic bomb survivors living in the United States, she wanted their voices to be represented as well.

Reynolds soon realized through her correspondence with Kuramoto that the U.S. hibakusha were very cautious about forming relationships with other organizations. She decided that she needed to meet with them personally, which she did the following spring during a fund-raising trip to California on behalf of the peace conference.

In Los Angeles, Reynolds visited Kaz Suyeishi, and with her help, met the leaders of the Japanese community. The Committee of Atomic Bomb Survivors was keeping its distance from Reynolds, but Suyeishi offered her a place to stay while she was in Los Angeles. During an evening at her home, Reynolds invited Suyeishi to attend the thirtieth anniversary conference at Wilmington College. She was about to accept, hesitantly, when the phone rang. Suyeishi couldn't understand the caller, and she handed the receiver to Reynolds.

It turned out the caller was Sanford Gottlieb, executive secretary of SANE, a well-established organization dedicated to banning nuclear weapons. Gottlieb was an old acquaintance of Reynolds' from the peace movement—they had demonstrated together in front of the White House—and their conversation progressed rapidly. Gottlieb described his plan to mark the thirtieth anniversary of the bombing of Hiroshima and Nagasaki with some special activity to promote the campaign against nuclear weapons. He too had heard about the U.S. hibakusha and had been given Suyeishi's telephone number.

Gottlieb wanted to know what the American hibakusha were thinking and what the goals of their organization were. Reynolds was able to give him that information, and she also told him that Suyeishi and Kuramoto would be attending the Wilmington conference from August 1 to 5.

"If they do, would it be possible for both of them to come to Washington to attend a press conference on August 6?" Gottlieb wanted to know.

Suyeishi had to be cautious. She didn't know anything about SANE, and she didn't want the atomic bomb survivors to be used for political purposes. She had been approached before by peace groups and anti-nuclear weapons organizations from Los Angeles and San Francisco, who wanted to use her name or the organization's. But she had always rejected these approaches because of the hibakusha's fear of getting involved in political activities.

Reynolds reassured Suyeishi that SANE was not affiliated to any political group and had a long history of educating the public about the dangers of nuclear weapons and the effects on radiation. When Suyeishi expressed concerns about her limited budget, which was barely enough for the Wilmington trip, Gottlieb promised that SANE would take care of the round-trip tickets from Wilmington to Washington and all expenses during their stay in Washington.

Reynolds's trip had proved instrumental in bringing the U.S. hibakusha in contact with the peace movement, and the presence of two U.S. atomic-bomb survivors at the Washington press conference was thus assured. Some hibakusha remained strongly opposed to this involvement, however. When a meeting of the southern California membership of CABSUS was held to authorize Suyeishi's trip, criticism was voiced by an engineer who worked in the defense industry. One of a small number of survivors who had earned an advanced university degree after the war, he threatened to quit the association if they endorsed the trip. Still, the endorsement was given.

At the peace seminar, Suyeishi and Kuramoto discussed the many problems that faced American hibakusha as both victims and citizens of the country that had perpetrated the atomic bombing. Their presence was of value not only because they were able to interact with atomic bomb survivors who came from Japan, but because they made a sympathetic impression on the 150 people who attended. Tōmin Harada from Hiroshima declared at the meeting that everybody on earth had become hibakusha through man-made radioactivity, and survivors of Hiroshima and Nagasaki were leading the campaign to abolish nuclear weapons.

But one participant noted how tentative the two representatives of CABSUS were about their involvement with the peace movement. "Since the peace movement is considered to be a 'red' activity in this country," they told him, "please do not make too much of our problems, because it would not help us." They were reluctant to get involved with a planned visit by the Japanese anti-A- and H-bomb organization to the United Nations.[2]

Suyeishi and Kuramoto flew to Washington early on August 6 to attend the SANE press conference, which was held at the National Press Club, near the White House. The largest room in the club had been reserved, but the press conference was moved to a smaller room at the last minute because Japanese Prime Minister Takeo Miki was speaking at the same time and the larger room was requisitioned for security purposes. As it happened, even the smaller room was too large for the meager turnout. Japanese reporters were busy covering the prime minister, and none of them showed up.

Ups and Downs

The previous evening, Ike Papas of *The CBS Evening News*, had "found" and interviewed hibakusha in California. Papas reported that the bill to aid American atomic bomb survivors was likely to be opposed by the Defense Department. Papas attended the SANE press conference, and his story was carried nationwide on CBS. Fifteen other reporters also covered the press conference, although only a few newspapers carried the emotional story of the two hibakusha, who spoke of their thirty years of fear of sickness and death.

Gottlieb later declared, "We missed the main target, but it was worthwhile because Mary McGrory of the *Washington Star* wrote about the U.S. hibakusha in her column."[3] Mary McGrory was syndicated in newspapers throughout the country and had a national reputation for integrity. Although she did not attend the press conference, she interviewed the two atomic bomb survivors at the SANE office, and her column, entitled, "Hiroshima Survivors: They Suffer 'Til Death," appeared on August 8.[4] It summarized the appeals of the two hibakusha and pointed out the irony of these survivors facing the possibility of death from radiation effects for the rest of their lives, while the nation's leaders talk about the feasibility and "rationality" of a "limited" nuclear war.

In McGrory's column, Kuramoto was quoted as saying, "They do not care about little human beings, and now the bombs are much worse. Why have No. 3? If they would come to Hiroshima and Nagasaki, all the leaders, American, Chinese, Russian, and see the people and what the bomb did to them, they would not talk that way."

Thirty years after the atomic bombing, the existence of hibakusha in the United States had begun to gain some attention. Yamada's survey of U.S. survivors, the visit of Barbara Reynolds, and the participation of Suyeishi and Kuramoto in the Wilmington and Washington events made more visible the picture of U.S. survivors and the America in which they lived.

In October 1975, the Japanese emperor made his first visit to the United States. Emperor Hirohito was once vilified as one of the three villains of World War II, along with Hitler and Mussolini, but now, nearing the fiftieth year of his reign, he was to be given a warm welcome.

Atomic-bomb survivors in America had no desire to raise objections to the visit, even though they knew that the Pacific War had been initiated with Hirohito's approval. The hibakusha were prepared to welcome the emperor in their own way, because they felt that his visit was good for promoting friendly relations between the United States and Japan. Wedged as they were between the two countries, they placed great value on good relations.

As the emperor's visit drew near, however, leaders of the local Japanese community in San Francisco delivered a notice to Kuramoto that read in part: "While His Majesty the Emperor is visiting the U.S., the hibakusha shall refrain from all activity. It would not be desirable to use the name of the Committee of Atomic Bomb Survivors as a welcoming organization."[5]

The emperor came and went. On a cold, rainy night in late October 1975, a quiet gathering of local atomic bomb survivors and their supporters, about twenty in all, took place at a church in the Nihonmachi section of San Francisco.[6]

These were hibakusha who had been refused permission to set up an information table for CABSUS at the spring Cherry Blossom Festival. They were the same people who had been asked not to engage in any activities during the visit of the emperor. The disappointment of having the medical aid bill voted down by the Senate Finance Committee in June was still fresh in their minds, along with the shock of being called "the enemy."

After everybody's spirits had been revived by a potluck dinner, Barbara Reynolds stood up and began to speak quietly about how she came to be involved in the hibakusha's problems. For most of those gathered there that night, even Barbara's simple English was hard to understand, so Kuramoto translated for them. She touched upon Yamada's survey and expressed her own opinion that the research and scientific investigations of the ABCC could not comprehend the feelings of the atomic bomb survivors. The ABCC had been saying that the effects of radiation were minimal. Softly, but with conviction, Barbara declared that the government had, in effect, been lying. She concluded by saying that when the time came, and a congressional hearing was held in Washington on the proposed bill to extend help to the hibakusha, all atomic bomb survivors in the United States should unite and present their case.

At a time when the morale of the survivors was so low, even a single friend was greatly valued. Encouraged by Reynolds's remarks, the meeting took up the revised survivors aid bill that was to be presented again in the state senate. "This organization has been in existence for three years already," one participant remarked. "We need some success." Even a compromise bill would be progress, and the discussion turned to the content of the bill. "What if we limit the bill to those who have U.S. citizenship?" one survivor proposed. "To limit the scope, let's exclude the children of survivors and stay away from the words 'Hiroshima' and 'Nagasaki,'" another suggested.

There was opposition from younger activists. "This bill will not pass anyway. Let us stick by our principles!" they argued. These young people, some of whom were said to be recent arrivals from Japan, called themselves the "Japanese Community Service." A few days prior to the meeting, a young pastor recounted, they had sent a letter to the church declaring that since the survivors had not tried to change the present social system, they were not interested in cooperating with the hibakusha organization. But it was unrealistic to expect the weak CABSUS organization to try to change the existing social system. It was clear that the survivors were not interested in politicking for the sake of politics.

In addition, the hibakusha organization was facing serious organizational problems. The Committee of Atomic Bomb Survivors in the United States had been functioning as two separate groups, the original southern California committee and the northern California committee, formed in 1974. But the southern California group was having trouble maintaining itself because of a lack of leadership, and it eventually merged with its northern counterpart in May 1976. Kanji Kuramoto was chosen as president of the united committee, and Kaz Suyeishi was named vice president. Kuramoto was not elected by a vote of the membership, however, and the merger did not necessarily solidify the organization. The appointment of Kuramoto, a northern Californian, sowed the seeds for later conflict within the organization.

The greatest fear of the American hibakusha was that they would be manipulated by a political interest group. Some six months after the gathering in San Francisco, in the spring of 1976, leaders of CABSUS were to lose many nights' sleep during the campaign over Proposition 15, an electoral initiative that engulfed the whole state of California.[7]

Proposition 15 called for guarantees of safety at nuclear power plants built in the state of California. The initiative called for one hundred percent compensation for damages caused by any accident in a nuclear power plant. (At the time, compensation was limited to a small portion of the damages.) In addition, all new construction of nuclear power plants would be halted if, within five years, two thirds of both the senate and the assembly could not be convinced of the safety of their operation and waste disposal. All plants then in operation would also gradually be taken out of service.

The utility companies opposed the proposition, which would have made it virtually impossible to operate a nuclear power plant in California. They also feared its influence would spread throughout the United States. Industry groups raised $4 million and, in league with labor unions and other powerful groups, mounted concerted opposition to Proposition 15.

Supporters of the initiative, primarily environmental groups like Friends of the Earth and the Sierra Club, launched a grassroots campaign throughout the state. Another Mother for Peace, headquartered in Beverly Hills, joined the effort to stop nuclear power in order to prevent another Hiroshima. The industry coalition used television ads and leaflets to attack the proposition, while the supporters campaigned house to house to convince voters to vote "yes."

One night close to election day, a stranger telephoned Kuramoto, who lived in Alameda, across the bay from San Francisco. The caller did not identify himself, except to say that he was opposed to Proposition 15, and he asked for Kuramoto's help. He promised financial assistance to the atomic bomb survivors, if they would cooperate with his side. "If your organization supports us, it will be a tremendous help," he said.

Support from CABSUS would have been of great value to the opponents of Proposition 15. An endorsement by the victims of the atomic bomb, people with first-hand experience of the damage caused by radiation, would have conveyed the impression that nuclear power plants were safe. Perhaps the caller knew that the hibakusha group was weak and in need of financial aid.

"I was shaken," Kuramoto said. The call totally ignored, and even went against, the feelings of the survivors. Kuramoto did not need to consult his organization. His answer was "no." Sure, they were in financial need, but victims of the atomic bomb could never support nuclear power.

He received similar telephone calls a few times after that, but the callers never divulged their names. Suyeishi, in southern California, also received anonymous calls. Her answer was also "no." She said that among hibakusha she was the one who most feared that their organization would be used for political purposes. For hibakusha who wanted only to live quietly in the United States, it was a huge effort just to ask for passage of a medical aid bill. To join a political group and thereby become the object of the anger and hatred of an opposing group was more than most members of the organization could endure.

The organizations supporting Proposition 15 did not offer the hibakusha financial help, but they did volunteer to help push the hibakusha bill through the state legislature. If necessary, they said, they would even go to Washington. If the atomic bomb survivors had expressed fear of radiation based on their own tragic experience, they would have been very persuasive. Kuramoto sympathized with the environmental organizations and he wished he could join the anti-nuclear power group. But the hibakusha organization had to say "no." Kuramoto and the other members of the organization believed that strict neutrality was the only way their weak organization could survive.

At the polls on June 8, Proposition 15 was defeated by a landslide of two to one. Two million voted "yes," but the opposition tallied four million. This did not mean, however, that two-thirds of the voters of California were convinced that nuclear power plants were safe. It merely meant that the industry-labor coalition had convinced voters that if Proposition

15 were passed, all nuclear plants in California would be closed, and California industry would be hard hit.

Two weeks after the referendum vote, the biannual national convention of the Japanese American Citizens League (JACL) was held in Sacramento.[8] The JACL, the only Japanese American organization of national scope, had a membership of 31,000, with 106 chapters around the country and an office in Washington with a full-time staff. The JACL was planning to adopt a resolution in support of the Roybal bill, which had been dormant for so long in Congress. The executive committee had made the decision to support the hibakusha bill the previous February.

After Suyeishi and Kuramoto attended the SANE press conference, Wayne Horiuchi, the Washington representative of JACL, together with Sanford Gottlieb and a staff member from Congressman Roybal's office, had visited members of the House Judiciary Committee to impress upon them the importance of the bill. But they found that some members did not even distinguish between Japanese Americans and Japanese nationals residing in the United States. Horiuchi reported this discouraging situation at the conference in Sacramento.

The feeling of the leadership had been that work on the bill should be put on hold until the vote on Proposition 15 in California. Now the proposition had been defeated and the political climate of California was temporarily calm. The JACL conference was held at an opportune time.

There were three major proposals on the agenda: (1) To ask for a presidential pardon for Iva Toguri, the Los Angeles-born Nisei who was known during World War II as "Tokyo Rose" and who had been convicted of treason; (2) To request compensation for damages incurred in the forced evacuation of the Japanese during World War II; and (3) To support a medical aid bill for atomic bomb survivors in the United States.

All three of the proposals were demands for restitution for past actions of the U.S. government. That the JACL belatedly decided to support Iva Toguri reflected the changing times. Previously, Toguri had been seen as a traitor, the shame of the Nisei. It had become clear over the years, however, that she had been made a scapegoat because of anti-Japanese sentiment that persisted into the postwar period. She was more properly seen as a victim, caught in the conflict between two countries, who had maintained her loyalty to the United States throughout.

As for compensation for damages resulting from the forced evacuation, the government had paid out small sums back in 1948. Issei who received these token payments had been dissatisfied, but they had said little at the time. The Nisei had tried to be "good" Americans and forget the bitter experience. That the issue of compensation was raised again thirty-four years after the relocation was a reflection of the heightened racial pride of young Japanese Americans of the third generation, Sansei.

The new demand covered not only material losses from the forced evacuation, but also psychological and cultural losses resulting from the long incarceration in the hot desert camps. Although these losses could never be truly compensated for by cash payments, an astronomical sum in damages was demanded to force the government to acknowledge that this type of injustice should never be repeated.

The request for aid to U.S. hibakusha was less assertive in its intent. It did not criticize the government for using the atomic bombs, and it did not seek compensation for all of the damages the hibakusha had suffered. The proposal sought to make the government aware that the hibakusha were being neglected, despite the fact that they were American citizens, and to ask for medical assistance. At the core of the proposal was the desire of the atomic bomb survivors to be recognized as fellow Americans, and the endorsement of this effort by their fellow Japanese Americans was of great significance.

When the proposal was presented, Thomas Noguchi, representing supporters of the hibakusha, provided background on the atomic bomb survivors and the efforts to obtain medical aid. Kanji Kuramoto delivered a passionate appeal, describing the plight of the survivors. Both of the speeches made a strong impression, and on June 24, the resolution to support the Roybal bill was unanimously adopted by the JACL. Horiuchi, the JACL Washington representative, promised lobbying efforts on behalf of the bill and expressed high hopes for its passage. The emotional Suyeishi spent the rest of the day moving through the hall thanking delegates, her eyes filled with tears.

Notes

1. This account is based on interviews with Barbara Reynolds and Kaz Suyeishi, April 3, 1978, Los Angeles. Also see a recent biography, Mizuhoko Kotani, *Hiroshima Junrei: Baabara Reinoruzu no shōgai* (Pilgrimage to Hiroshima: The Life of Barbara Reynolds) (Tokyo: Chikuma Shobō, 1995).

2. Accounts of the conference were provided by Reynolds and Tōmin Harada; see also Harada, *Hiroshima no gekai no kaisō* (Reminiscences of a Hiroshima Surgeon) (Tokyo: Miraisha, 1977), p. 260.

3. Interview with Sanford Gottlieb, April 11, 1978, Washington, D.C.

4. Mary McGrory, "Hiroshima Survivors: They Suffer 'Til Death," *Washington Star*, August 8, 1975.

5. This account was provided by Kanji Kuramoto.

6. Accounts of the meeting of hibakusha are based on the author's personal observation.

7. On Proposition 15, accounts were provided by Kuramoto and Suyeishi; also the author covered the campaign over the proposition for the Japanese press. See Rinjiro Sodei, *Hankaku no Amerika* (Anti-Nuke America) (Tokyo: Ushio Shuppansha, 1982), pp. 186–201.

8. The account of the 1976 JACL convention is based on the author's own observation.

fifteen

A Medical Team Comes and Goes

Immediately after Kaz Suyeishi returned from the 1976 JACL convention in Sacramento, she began to make preparations for a trip to Japan with her daughter Christine.[1] Kaz had not visited her family in Hiroshima for six years, and with her mother nearing ninety years old, Kaz wanted to show her how Christine had grown into a mature woman. Her husband was supportive and urged Kaz not to make a hurried trip, but to enjoy a good rest in Japan.

In addition to the family visit, Kaz had another very important mission in mind. In her official capacity as vice president of CABSUS, she hoped to find out what chance there might be of having a group of medical specialists visit the United States.

The reader will recall that when the U.S. hibakusha group was in its infancy, Tomoe Okai, who was then its leader, had repeatedly asked Hiroshima Mayor Setsuo Yamada to send a team of medical specialists to the United States. The mission had never materialized, despite an announcement in October 1971 that the mayor was hoping to arrange a medical task force to visit the United States in March of the following year. The task force was abandoned, and instead Dr. Hiroshi Maki of the ABCC conducted a group examination in Los Angeles in August 1972.

Despite the limited scope of Maki's examination, it had spurred the AEC to organize a survey of U.S. hibakusha and indirectly led to the first public recognition of the hibakusha in America. The survivors continued to believe that thorough examinations by a medical task force skilled at diagnosing the effects of the atomic bomb would do a great deal to alleviate their anxiety.

A Medical Team Comes and Goes

The Japanese government appeared reluctant to send a medical team, however, perhaps out of deference to the U.S. government. In a letter dated August 7, 1974, Kanji Kuramoto, in the name of CABSUS, had asked the Japanese government about the possibility of a medical task force and the issuance of hibakusha identification cards to those who made their homes in the United States. An answer arrived almost a year later, on June 3, 1975, via the Japanese consul general in San Francisco.[2] The letter said the Japanese government would respond with flexibility to the request to issue certificates to atomic bomb survivors in America, but that it would be difficult to comply with the request for a medical task force, because of medical licensing agreements between the United States and Japan.

Kaz Suyeishi refused to let the matter drop. A-bomb survivors in the United States would need the help of the city of Hiroshima as well as the backing of the Ministry of Health and Welfare, and Suyeishi took upon herself the task of dealing with these two government bodies. It was clear that the hibakusha group could not provide financial support for her efforts. She realized she would have to proceed with support from her understanding husband, what little money she could afford herself, and whatever time she could spare from her visit to her mother.

The record of her persistent efforts, from the time she arrived in Japan on July 3 until she returned to Los Angeles on November 15, would comprise a whole chapter in itself. She made it her first priority to establish contact with the Ministry of Health and Welfare in Tokyo, which she considered important enough to postpone her visit to Hiroshima. Her efforts bore fruit. Four days after her arrival in Japan, an interview was arranged with the minister of health and welfare, followed by another meeting two days later. Suyeishi requested Japanese government aid to the U.S. hibakusha and elaborated on the anguish and suffering endured by her fellow survivors. The news media—the Kyodo News Agency and the *Chūgoku Shimbun*, in particular—were responsive to her plea and reported her movements in Japan sympathetically. One newspaper headlined a report: "Heartless Homeland," referring to the United States. Deep within, Suyeishi did not think that Japan—the adoptive "homeland" of her youth—would turn its back on the hibakusha.[3]

Soon after Suyeishi arrived in Hiroshima, her mother was injured in an accident and hospitalized. Suyeishi's work nevertheless intensified. She went to city hall to carry her plea to Mayor Takeshi Araki. On August 6, she participated in the anniversary services at Peace Park. On the 9th, she traveled to Nagasaki, where she met with the mayor to enlist his help, informing him that there were eleven hibakusha from Nagasaki residing in the United States. When she returned to Hiroshima, an urgent call was waiting for her from the mayor's office. The mayor wanted to see her and to assure her that he was willing to intervene on her behalf with the Ministry of Health and Welfare.

With support coming from all quarters, Kaz Suyeishi's efforts were beginning to bring results. On August 26, the minister of health and welfare invited Suyeishi to his office to inform her, "The Japanese government has decided to send a medical team specializing in A-bomb symptoms to the United States and has liberalized the conditions for issuing hibakusha ID cards to those residing in the U.S." The *Chūgoku Shimbun* of August 27 carried this report:

> Because of American laws governing the practice of medicine, the medical team being sent to the United States will be in that country on a research project of the Radiation Effects Research Foundation (formerly the ABCC). The size of the team, the time of its departure, and the duration of its visit will be determined after consultation with appropriate counterparts in the United States government, but the Ministry of Health and Welfare is contemplating sending two doctors, an internist and a surgeon, for a period of about a month.[4]

The newspaper report included the following comment by the chief of the Ministry's Public Health Bureau: "We would like to have a staff doctor from the Atomic Bomb Hospital join the team. We could send the team before the end of this year, assuming preparations are completed on the other side."

Regarding hibakusha ID cards, it was explained that if an application were submitted prior to an applicant's coming to Japan, the card would be ready when the applicant arrived. This procedure relieved overseas hibakusha of the requirement to produce two witnesses to establish that an applicant was a hibakusha, a condition that was even difficult for hibakusha in Japan to fulfill. Kaz Suyeishi applied for her ID card on August 30 at the Hiroshima city hall. She became the eighth A-bomb survivor residing in the United States to receive a hibakusha ID.

The decision had been made to send medical specialists to the United States, but this was just the beginning of the task. First, there was the problem of how the medical team would be received when it arrived. CABSUS was loosely organized and lacked influence. The members were handicapped by the language barrier, a lack of funds, and the fact that most of the survivors were middle-aged or older women. Few paid their $10 or $15 annual dues regularly, and very few contributions were ever received.[5]

Which hospital would provide facilities to hold the examinations? Who would pay the costs? How would the clerical work be handled? The U.S. hibakusha group was elated at the news of the medical team's prospective visit, but they were at a loss as to how to cope with the arriving doctors. The organization did not even have an office of its own. Like many

A Medical Team Comes and Goes

Japanese organizations, it had no clear line of authority or responsibility. Kaz Suyeishi was in a quandary. Someone had to be in the driver's seat, but she knew that anyone who stepped forward would be subject to criticism and personal attack.

At about this time, while Suyeishi was still in Japan, plans were set for the mayors of Hiroshima and Nagasaki to visit the United Nations in November. Their itinerary included a visit to California on the way back to Japan. Suyeishi thought that she should return to California before the mayors arrived, so that she could solicit their help in making arrangements for receiving the Japanese medical team. Before her departure, she was invited to dinner by Gorō Ōuchi, president of the Hiroshima Prefectural Medical Association. He told her that he would be accompanying the mayor of Hiroshima on the forthcoming trip, and while in California he intended to propose a sister relationship between his organization and the Los Angeles County Medical Association. This struck Suyeishi as a surprising idea, but one that would be a great help in paving the way for the Japanese medical team. It was evident that Ōuchi was already aware that the chair of the county medical association and his counterpart in the Southern California Japanese American Medical Association were strong supporters of the hibakusha medical aid bill.

Satisfied that she had accomplished what she had set out to do, Suyeishi returned home to Los Angeles after an absence of four months. What she found upon her return was bewilderment and confusion among her fellow A-bomb survivors, who had been unable to agree on plans for the Japanese medical team. The whole idea had become a burden rather than a blessing to the poorly organized group. CABSUS President Kuramoto eventually concluded that in order to pull his organization together, he would use his Christmas vacation to make a trip to Japan, confer with authorities at the Ministry of Health and Welfare and the RERF to find out their plans regarding the visit and, on that basis, formulate a concrete plan for receiving the medical team.

On December 26, Kuramoto arrived in Hiroshima. He was met at the station by a reporter from the *Chūgoku Shimbun*, to whom he said, "The hibakusha in the United States have sincere hopes that the proposed medical team will be of a caliber capable of diagnosing their actual condition, that they will prepare a report of their findings for submission to the U.S. Congress, and that this entire undertaking will not start out with a bang and end up in a fizzle." When he added that the dispatch of a medical team should not be just a one-shot affair but a long-range program, he was voicing the hopes of all the A-bomb survivors in the United States.

The following day, Kuramoto visited the mayor's office, after which he met with the press and gave them an account of the conditions of the hibakusha in the United States. He said, "We are pressing the U.S. Congress for passage of a bill providing financial support for A-bomb survivors' medical needs. If a medical care system can be put together first, we will have overcome the first hurdle." With the Japanese government reaching out to provide medical aid for hibakusha in the United States, he believed, the U.S. government would have to take action.

That same day, Dr. Ōuchi informed Kuramoto that the Los Angeles County Medical Association had approved a sister-organization relationship with the Hiroshima Prefectural Medical Association. Ōuchi told the press, "With this relationship established, the groundwork has been laid for the visit,"[6] and this turned out to be the case.

During Kuramoto's stay in Hiroshima, a troubling development took place. The minister of health and welfare had announced that the medical team would go to the United States as a function of the Radiation Effects Research Foundation (RERF), which was now under the jurisdiction of the ministry (in conjunction with the U.S. ERDA, the successor to the AEC). The role of chairman of the foundation board alternated between the Japanese and the Americans. The first chairman was Hisao Yamashita of Japan. When Kuramoto visited the RERF to confer with officials about the visit of the medical team, he was told that the foundation had received no information about the mission from the Ministry of Health and Welfare. Masuo Takabe, one of the directors of the RERF, met with Kuramoto and told him definitively that the United States was outside of the territory covered by the RERF's activities, so they would not be able to do anything to help the A-bomb survivors in the United States. Takabe produced an English version of the RERF articles of incorporation and read aloud: "Any activities rendered by this corporation would be restricted to within the boundaries of the national territory of Japan. Various researches and investigations listed hereunder shall be performed and executed for the people residing within Japan."[7] Takabe concluded that it would thus be difficult to provide medical examinations for the U.S. hibakusha, but they would think about it.

These statements came as a shock to Kuramoto. The U.S. government had invested huge sums of money in the ABCC (now the RERF), but it seemed that the benefits of decades of research would not be available to the A-bomb survivors in the U.S. This was incomprehensible to Kuramoto, since the American hibakusha were taxpayers and felt entitled to benefit from their government's activities. Eventually a solution was worked out, however.

In early spring of the following year, the *Chūgoku Shimbun* reported that a medical team would be sent to the United States under the joint auspices of the RERF and the Hiroshima Prefectural Medical Association.[8] Although the article was prefaced by the word "unofficially," an agreement had been reached between the chief of the Public Health Bureau of the Ministry of Health and Welfare and the Biochemical and Environmental Research Section of the ERDA.

The two sponsoring organizations would assume responsibility for the budget of five million yen ($25,000), and the medical team was announced as including Dr. Taiji Okada, a member of the board of directors of the Hiroshima Prefectural Medical Association; Assistant Professor Michihiro Miyanishi, of the Hiroshima University Medical School, who also held a position of clinical consultant to the RERF; and Hiroaki Yamada, chief of the survey section in the Epidemiology and Statistics Department at the RERF, who had conducted the earlier survey of hibakusha in the United States. The trip would last three weeks, and the doctors' activity in the United States was referred to as "health consultation" rather than medical examination.

The visiting Japanese medical team, headed by Dr. Okada, arrived in the United States as promised and conducted examinations and health consultations, virtually without rest, from March 29 to April 8 in Los Angeles and from April 9 to April 15 in San Francisco, Sacramento, and San Jose. In Los Angeles, the local Japanese American medical association provided extensive cooperation and the facilities of City View Hospital were made available. Sixteen Japanese American doctors from the area took turns standing by as consultants, while their wives served as nurse's aides. All of this was done on a volunteer basis. Because of the tremendous help, it was possible for the medical team to go beyond oral histories and conduct thorough medical examinations on each of the thirty-four hibakusha who showed up. The examinations included X-ray tests, electrocardiograms, and twenty-four different blood tests. The cost, excluding doctors' fees, was estimated at about $2,000 per person. The only compensation Okada and Miyanishi received was money for travel expenses. Their services were rendered gratis. Okada even paid out of his own pocket to hire a substitute doctor to run his clinic while he was abroad.

The thoroughness of the examinations in Los Angeles satisfied Thomas Noguchi, who hoped to obtain data on the U.S. A-bomb survivors to promote the medical aid bill in Congress. At the same time, the critical role played by the Japanese American Medical Association helped to dispel long-standing tensions between the hibakusha and the association, which were based on the hibakushas' sense that Japanese American doctors did not seem to understand or have compassion for the A-bomb survivors.

In northern California, the work did not proceed as smoothly, mainly because there was no comparable medical association to provide assistance. Japanese American doctors helped out individually, but the team could do no more than conduct medical histories. In San Francisco, a total of thirty-six hibakusha gathered to be examined in a meeting hall of the JACL national headquarters. In Sacramento, twenty-one survivors came to the Buddhist Church, and in San Jose, eighteen assembled for the examination at the office of a local Japanese American doctor.[9]

The conclusion of the visiting doctors, after examining 123 U.S. hibakusha, was that their suffering was primarily psychological rather than physical. The terrifying experience of the A-bomb had left an indelible mark on their minds, which had been exacerbated, in the United States, by the difficulty of the language barrier and cultural differences. American doctors had made little effort to understand the psychological and emotional aspects of the hibakusha's suffering. Now a group of doctors had come from Hiroshima who spoke their language and examined them with compassion and understanding. The net result for the A-bomb survivors was the attainment of a certain peace of mind. It was the same in southern California and in northern California. The hibakusha who gathered all wore intensely worried looks before the examination but emerged afterward looking relaxed and peaceful.

From all accounts, the medical mission accomplished its purpose, and those who were examined were satisfied. At the JACL meeting room in San Francisco, for example, the survivors assembled long before the time of their appointments and talked among themselves as if they were not aware of the passage of time. Kuniko Jenkins, whose lungs were seriously damaged by the aftereffects of the bomb, moved busily among her fellow hibakusha, serving tea and cookies, and seemed to enjoy what she was doing immensely.

Needless to say, there were some who refused to participate on the grounds that no follow-up to the examinations was contemplated. The examination, they felt, would only mark them as hibakusha and result in increases in their insurance premiums. These people would not be able to set their minds at rest unless a medical aid bill was passed that could dispel their fears of excessive medical expenses.

Ironically, the medical examinations were an important step toward that goal. The official medical report revealed one case of uterine cancer and one case of heart disease in Los Angeles. No other cases of abnormality that could be traced to atomic bomb effects were discovered. "You should not worry," Dr. Okada reassured the hibakusha, "You are all well enough to forget the past." Professor Miyanishi was somewhat more cautious. "We cannot say that this group represents a fair sample of the whole population of hibakusha," he declared, "because it has not been possible to trace the total group."

On the evening news the day the examinations were completed in San Francisco, Okada was interviewed by local TV and said, "There seem to be no serious medical problems with the A-bomb survivors here in the United States." Kuniko Jenkins' husband, who was watching the newscast, cried, "How can he say there are no problems!" He turned to his wife with tears in his eyes. "We have a living example of those problems right here in our home."[10]

～　　　～　　　～

The "forgotten Americans" had finally succeeded in gaining the attention of the Japanese atomic-bomb specialists, but the U.S. hibakusha continued to be a low priority as far as Washington was concerned. A well-informed observer commented that "Washington decided more than a year ago that the A-bomb survivors in the U.S. are not of interest [in terms of the pathology of radiation disease]. The government has no intention of making them the object of research." A San Francisco physician who had previously worked for the ABCC told Kuramoto something along the same lines: "The total number of A-bomb survivors living in the U.S. is too small and the estimated quantitative radiation absorption of each one is too small to warrant objective medical research. There are only about five individuals who received radiation of 100 rads or more, which is considered to be the minimum level capable of inducing a cancerous condition."[11]

John Gofman, the radiation specialist who was then professor emeritus at the University of California, had this comment: "It isn't that these scientists are not interested in the U.S. A-bomb survivors. They do not wish to be interested."[12] Kuramoto's efforts to gain support from U.S. scientists only reinforced this impression. After the Japanese medical team returned to Hiroshima, Kuramoto wrote to the National Academy of Sciences, the organization that was instrumental in establishing the ABCC, and asked for the names of medical professionals who had been involved in ABCC activities over the years. A list of about thirty names was sent to him. He had written a letter to President Jimmy Carter, describing the situation of A-bomb survivors in the United States and requesting the president's help. He made copies of this letter and sent it to the thirty doctors and scientists to enlist their support. One doctor sent a check for $30 to help the cause. The rest did not even respond.

The U.S. hibakusha had no recourse but to continue their efforts and hope for the best. Their immediate goal was to find a way to obtain regular visits to the United States by the Japanese doctors. But to realize this, it would first be necessary to pass the medical aid bill that had been stalled in Congress since 1972. Renewed hope for the bill was sparked by an editorial in support of the Roybal bill that ran in the *Los Angeles Times* on August 7, 1977. The editorial pointed out that the Japanese government provided medical assistance to atomic-bomb survivors in Japan, and argued that the U.S. government should do the same for hibakusha in the United States.[13]

Encouraged by this turn of events, Kuramoto wrote another letter to President Carter. He received a printed postcard acknowledgment. In January 1978, Kuramoto sent a third letter of appeal to President Carter. This time he attached the medical report prepared by the Japanese medical team and requested government support for continuing visits by Japanese medical teams to the United States. It concluded with the following words. "I am appealing to you to open your heart to aid these people in the spirit of true love."

A reply dated February 8 came from an unexpected quarter: the Japan desk at the State Department. The letter, from Edward M. Featherstone, began, "The White House has asked me to respond to your letters to President Carter." Kuramoto wondered if Carter believed that the A-bomb survivors in the United States were Japanese nationals. Reading on, he soon felt that a thick wall had been erected before him. Featherstone wrote,

> It has been the long-standing policy of the United States government, however, not to pay claims, even on an ex-gratia basis, arising out of the lawful conduct of military activities by U.S. forces in wartime. This policy is based on such considerations as the absence of any legal liability and difficulties in locating, singling out and determining the relationship of the A-bomb experience to current health problems. Additionally, the very great length of time which has passed since the bombing would cause practical difficulties for any investigations.
>
> In accordance with this long-standing policy, the United States has not, as you know, been directly involved in the treatment of those who were affected by the atomic bombs. This work has been primarily carried out by the Japanese government and Japanese medical institutions.

In essence, the U.S. government would not take responsibility for the A-bomb victims, on the grounds that the dropping of the A-bombs was a legal act of war. If any medical aid was to be given, the letter implied, it would have to come from private sources or from welfare agencies, or through passage of the Roybal bill in Congress. This was the lone ray of hope in the letter, but Kuramoto remained pessimistic. He feared that even if the bill was approved by Congress, Carter would surely veto it.

A matter of even greater concern, however, was that the U.S. government seemed to think of the U.S. A-bomb survivors as Japanese nationals. Why else would the response come from the Japan desk at the State Department? Kuramoto felt that this matter was beyond his ability to deal with. He immediately got in touch with Harry I. Takagi, then acting as the Washington representative of the JACL. Takagi wrote a letter of protest to the White House, informing the president that "Mr. Kuramoto is dissatisfied with Mr. Featherstone's reply, because (1) Mr. Featherstone showed no recognition of the fact that he was dealing with an American citizen, and (2) Mr. Featherstone's office appears to be one which deals principally with the affairs of Japanese in Japan, rather than with the rights of *American citizens of Japanese ancestry.* Mr. Kuramoto is not alone in his feelings; we who are loyal American citizens wish to be so recognized and treated as such, and not confused with citizens of a foreign country."

Takagi's letter also mentioned that the Japanese government had initiated a law to provide medical treatment free of charge to the atomic-bomb survivors and continued, "If the Japanese government has seen fit to do this on a humanitarian basis, can your administration, with its emphasis on human rights, do any less?"

"If it is considered necessary to refer this letter to another office," the letter concluded, "please do *not* refer it to the State Department Office of Japanese Affairs."[14]

The White House never replied.

Notes

1. Interview with Kaz Suyeishi, April 3, 1978, Los Angeles.
2. Letters exchanged between Kanji Kuramoto and the Japanese government are in Kuramoto's personal papers (hereafter cited as "the Kuramoto files").
3. Suyeishi's letter to the author, December 1, 1976.
4. *Chūgoku Shimbun,* August 27, 1976.
5. Interview with Kuramoto.
6. *Chūgoku Shimbun,* December 28, 1976.
7. Kuramoto files.
8. *Chūgoku Shimbun,* March 9, 1977.
9. The overall view of the Japanese doctors' first visit was obtained by on-the-spot interviews with Drs. Okada and Miyanishi while the author was observing the medical checkup of hibakusha in San Francisco.
10. Jenkins's reaction was conveyed to the author by Kuramoto.
11. Conversation recounted by Kuramoto.
12. John Gofman interview.
13. *Los Angeles Times*, editorial, August 7, 1977.
14. Kuramoto's letters to President Carter, the reply from the State Department, and Tagaki's protest letter to the White House are all in the Kuramoto files.

sixteen

Washington Comes to Los Angeles

The letter from the State Department indicated to the hibakusha that there was little prospect that the White House would provide support for their cause. Their only remaining hope was the medical aid bill that remained stalled in Congress.[1]

There were some new grounds for hope on this front. In 1974, Norman Y. Mineta, who had been mayor of San Jose, California, was elected to the House of Representatives, as the first Nisei congressman from the mainland United States. He immediately became a co-sponsor of the bill that originally was introduced by Congressman Edward Roybal, and with his sponsorship the Japanese American Citizens League (JACL) had a new incentive to support passage of the bill. In the summer of 1976, however, the entire JACL staff was replaced with new executives. Even the Washington representative, Wayne Horiuchi, was no longer functioning in that capacity. JACL support would have to be rebuilt from scratch.

Another encouraging development occurred, however, when Congressman George E. Danielson of San Gabriel, a suburb of Los Angeles, assumed the chairmanship of the Subcommittee on Administrative Law and Governmental Relations of the Judiciary Committee, which had responsibility for the bill. Danielson's constituency included a sizable number of Japanese Americans, and it was expected that he would give strong support to the bill. It was through Danielson's efforts that the first congressional hearing on the hibakusha aid bill was finally scheduled for Los Angeles on March 31, 1978, nearly six years after the bill was first introduced.

There had previously been some talk of holding a congressional hearing on the bill, but nothing had ever come of it. Supporters of the bill feared that even if a hearing were held it would certainly be in Washington, which would have made it extremely difficult to send any witnesses. When these supporters heard that the hearing was to be held in Los Angeles, they were overjoyed. For the first time in history, the U.S. Congress was going to listen to what the American hibakusha had to say, and Washington was coming to southern California, where the majority of the survivors lived.

Southern California does not usually get much rain, but that year it had rained quite heavily. In February, the rains had caused landslides in some parts of southern California. On the eve of the congressional hearing, rain fell all night, and when the hearings opened on the last day of March, the skies were cloudy over Los Angeles.

The hearing was held in the L.A. County Building, on a hill just to the east of Little Tokyo. The seating capacity of the hearing chamber was over seven hundred. The hibakusha and their friends were concerned about the turnout, because the number of people who come to a hearing is generally considered to be an indication of the level of interest among the general public. The problem of the hibakusha in the United States was not fully understood even within the Japanese American community, let alone by the American public at large. It was not anticipated that there would be many non-Japanese Americans at the hearing.

A press release had been sent to most of the local newspapers, but not a line was reported in either the *Los Angeles Times* or the *Herald Examiner*. Publicity among Japanese Americans, however, was quite intensive through Japanese vernacular newspapers and radio programs. The hibakusha were indeed of concern to many elderly Issei and Kibei Nisei, but it remained to be seen how many would actually show up at the hearing.

Well before ten o'clock, when the session was to begin, the doors to the hearing room were opened. Large numbers of concerned Japanese Americans stepped out of chartered buses and entered the building. Kaz Suyeishi, vice president of the survivors' organization, was overwhelmed at the sight, and could not hold back her tears. She had visited the Little Tokyo Tower retirement home and many Japanese American churches to ask their residents and members to come out to show their support. But she did so knowing that even if Issei were asked personally to come, and even if they were retired and had plenty of time, it would still take a great deal of initiative to come to the county building, which was essentially a white institution, and listen to a plan for political solutions to a problem from which they were far removed. Besides, they would not be able to understand most of the proceedings. Only those with an extraordinary level of concern for the survivors would be willing to sit through the hours of complicated discussion. Knowing this, Suyeishi was astounded by the turnout. One elderly Issei told her, "I don't think I'll be able to understand anything at all, but I'll be there because I don't want you to be disappointed."

It did not seem, however, that there were very many hibakusha present, except for those who were scheduled to testify and members of the committee, including the founding president, Tomoe Okai, and Suyeishi. Work was the main excuse given for the low hibakusha turnout. To a Japanese observer, it seemed a very American excuse, although, of course, this *was* the United States. It was not to be expected that the front rows would be filled with hibakusha and their supporters wearing headbands and with their sleeves rolled up, as would have been the case in Japan.

By ten o'clock, the room was half-filled. Almost all of those who attended were of Japanese descent, mostly gray-haired. They were saying to each other, "It's been quite a while since the last time," a reference to the occasion nine years before, when a public hearing was held in the same room on Thomas Noguchi's appeal for reinstatement to his office of county coroner. At that time, supporters argued convincingly on Noguchi's behalf and succeeded in having the county rescind his dismissal. Partially in return for the support of the Japanese American community, Noguchi had worked tirelessly to strengthen the hibakusha organization, arranged the support of the county medical association, and helped to draft the medical bill that was the subject of the hearing. Now Noguchi was on the platform with the congressmen, appearing as a key witness.

Congressman Danielson chaired the hearing. The hibakusha bill was referred to his subcommittee because it involved federal government payment of compensation. In Congress, chairpersons enjoy tremendous power, since they decide which bills are taken up for discussion. In each session of Congress, between twenty and thirty thousand bills are introduced, many of which die without coming to hearing. The hibakusha aid bill had been shelved three times in the past, and it was thanks to Danielson that it had finally come as far as a congressional hearing. Danielson's district included the Monterey Park area, where an increasing number of affluent Japanese Americans had moved from Los Angeles. There were at least fifty thousand Japanese Americans in his constituency.

It would be a mistake, however, to think that Danielson's support of the bill was motivated only by election-year politics. The hibakusha bill, now called HR 5150, was unpopular because it required compensation by the government, and it concerned only a limited number of people within the Japanese American minority. There is little doubt that Danielson understood and responded to the humanitarian aspects of the bill.

Before Danielson assumed the chairmanship two years earlier, his position had been held by Congressman Walter Flowers of Alabama. As a member of the Judiciary Committee, Flowers had stood firm for the impeachment of President Nixon in the Watergate case, but he could not escape the conservative constituency he represented. The strongest opposition to the hibakusha bill was reportedly coming from the South, and Flowers was a veteran congressman from Alabama. It was fortuitous, then, that Danielson had taken over chairmanship of the subcommittee.

There were seven members of the subcommittee, but only Danielson was present that day. On the platform with Danielson were the co-sponsors of the bill, Congressmen Edward Roybal and Norman Mineta. Danielson, empowered to do so by his office, allowed them to question the witnesses. All three were Democrats from California, and because they were all in favor of the bill from the outset, the hearing was conducted in a relaxed and congenial atmosphere. Also on the platform, as a legal expert in this area of the law, was the counsel to the Judiciary Committee, William P. Shattuck.

Seated in the front row were Kuniko Jenkins and her parents. Jenkins had experienced the atomic bomb explosion in Hiroshima at the age of nineteen, while she was on duty as an internship nurse at an army hospital in Hiroshima. She bore facial scars, a vivid and constant reminder of the tragic experience she went through. She was married to an American veteran who worked in a military medical institution. She had faced death several times in the past due to the aftereffects of exposure to radiation. Her lungs had deteriorated to the point that she required six inhalations of oxygen daily. Yet she came all the way from San Francisco to testify. No airline would allow her to carry her oxygen tank on board, and her husband could not get off work to drive her, but she was so determined to appear that she drove all the way to Los Angeles herself, bringing her elderly parents in the back seat for support. It took eight hours for a healthy person to make the drive, but Jenkins and her parents stayed overnight in Fresno and, stopping many times along the way, took two whole days to drive to Los Angeles. Her doctors would not give her permission to speak at the hearing, so when she took the witness stand, her testimony had to be read by another person.

On the wall behind Danielson and facing the audience was a large white banner with a design of the mushroom cloud in purple and the caduceus of Hermes representing medical treatment. Ten feet wide and fourteen feet long, it was the creation of members of a support organization. It symbolized the need for health care for survivors of the atomic bomb explosion.

The chairman, a warm-hearted man in his early sixties, opened the hearing by stating:

> We all know that in August of 1945 there were atomic bombs dropped over Hiroshima and Nagasaki. One thing that many of us do not know in the United States is that there were many American citizens present there at the time. . . . There are probably eight hundred to one thousand survivors present here in the United States. We don't know the exact number. We understand that approximately 150 of them still require continuous medical and related treatment. We don't know exactly how many. . . . The idea of the bill is to provide that for U.S. citizens and permanent residents who are suffering from radiation sickness of one type or another, . . . that the U.S. government, in the future, for future incurred expenses, pick up the cost of paying for that type of treatment. . . .
>
> There is one other tremendously important aspect of the bill, from the scientific point of view . . . there really is no one central bank of information on how radiation sickness affects people, how it should be treated, where, when, and so forth; and if this bill did become a law, it might make it possible to accumulate such a bank of information which could be held in reserve against the day that conceivably it might be needed again. Hopefully, we never will, but. . .

The American government has always maintained that dropping the atomic bombs during wartime was a legal act of war. The American hibakusha thus did not claim damages caused by the bombing of Hiroshima and Nagasaki. They were simply requesting financial aid for the medical treatment of symptoms caused by late effects of radiation that might strike them as they grow older. This was a modest request, considering all they had endured through the years. The hearing was an attempt to establish, through the testimony of survivors and their supporters, that the government had a future responsibility to assist in their health care.

The first witness was Thomas Noguchi. His testimony provided historical and medical observations about the problems facing survivors of the atomic bombings, and a powerful appeal for passage of the bill. It was an eloquent summation of the plight of the U.S. hibakusha. He concluded by stating, "We are not talking about health care for Japanese or for foreign subjects. We are talking about health care for Americans—the quiet Americans who have suffered for many years from a variety of conditions caused by the atomic bombings. I do not need to belabor this. The logic is very clear. It is the only purely humanitarian standpoint."

Before questions were asked of Noguchi, Mervyn Dymally, the lieutenant governor of California, took the witness stand. He spoke briefly about the State Senate hearing he had chaired and about the dilemma the survivors face because, as working people, they are not poor enough to qualify for state medical aid, and their physical symptoms are so subtle that they cannot be identified easily as victims of radiation-related diseases.

Dymally spent the remainder of his time reading a letter from one of the survivors. The letter concluded: "Our race and language may be different, but our warm feelings will not change. I pray and pray and tears flow down my cheeks when I think how all our group are still suffering all these years without any cure." Dymally added, "This was the most eloquent letter I have ever received in the years I have been in politics."

Congressman Roybal asked Noguchi about Japanese legislation on health care for hibakusha in Japan and about the budget necessary for implementation of the Roybal bill. "We hope that perhaps a $250,000 budget allocation for the first year might be adequate to cover the initial establishment of this program," Noguchi replied. That amount is a small fraction of the cost of one of the many weapons in the American nuclear stockpile.

Throughout, Roybal's aim was to clarify the content of the bill by his questions rather than to interrogate the witnesses. When the minutes of the hearing were published and made available to other members of Congress, they would provide the necessary background on the bill.

The next witness was Dr. Samuel Horowitz, the former president of the ten-thousand member Los Angeles County Medical Association. Horowitz told the hearing, "I am particularly proud to announce to you that not only the California Medical Association but also the American Medical Association have both endorsed [HR 5150]." The proposal to endorse the hibakusha bill had been tabled at the AMA convention in San Francisco in June 1977, but it was finally adopted by the board of directors at a meeting in Chicago at the end of that year. Horowitz's success in winning the support of the conservative AMA was quite significant. The AMA endorsement was later cited by every congressman who supported the bill. Horowitz concluded his testimony by suggesting an addition to the bill to authorize annual physical check-ups for all hibakusha covered by the bill.

It was almost an hour since the hearing began. Kuniko Jenkins was quietly inhaling her oxygen. It was time for testimony from the survivors themselves. The first was Judy Enseki. Anyone who had seen Enseki just a few years before would have been shocked when she stepped to the podium. When I interviewed her in the summer of 1976, she did not look or behave at all like a hibakusha. Now she was gaunt, and her pallor was disturbing.

From the witness stand, Enseki testified about her present condition. "I am currently under treatment for anemia and thyroid problems," she stated. "The group medical policy at my place of work does not cover the cost of treatment, since this type of anemia is treated with vitamin-type injections and medication, and the policy does not cover preventive care."

Responding to questions, Enseki testified that her health care expenses at the time amounted to $700 a year, and neither her present group insurance nor her previous Blue Cross policy covered the treatment for this type of blood disease. (In 1980, Judy Enseki, who had been one of the healthiest and most well-adjusted of survivors, was hospitalized with cancer. She died on August 21, 1980.)

Judy Enseki's case provided vivid evidence of the hazards and uncertainties the survivors face as they grow older. Many people who were exposed to radiation from the atomic bomb in their youth led healthy lives until they reached middle age. Then, all of a sudden, they began to develop various symptoms, including cancer, and there was no way of telling whether these were naturally occurring diseases or could be traced back to the bomb. Karl Z. Morgan, then a professor at Georgia Institute of Technology, served for thirty years as the head of the Health Physics Division of Oak Ridge National Laboratory, the leading center for research on radiation effects on the human body. "No matter how small the amount of radiation to which one is exposed," he said in a telephone interview, "one cannot disregard its potential hazard. The fact that a survivor of the atomic bomb has lived so far without any abnormal symptoms does not mean they will be free from any danger in the future."[2]

It was Kuniko Jenkins's turn. She walked slowly past her concerned parents to the witness podium, carrying her oxygen tank. The entire hearing room was hushed. Enseki read Jenkins' testimony for her: "In some ways, I am very fortunate. I have been married to a wonderful man, who accepted me even knowing that I had survived the atomic bombing of Hiroshima. My husband is retired from the military, and I have been able to receive medical treatment without undergoing the extreme financial suffering that many survivors in the U.S. have had to face." She described vividly the scene of the atomic bomb explosion, her miraculous survival, and the days following the tragedy, then closed her testimony by saying: "We ask for your help in supporting HR 5150, and we hope, above all, for world peace." Her aged parents listened to her attentively with their eyes closed. They may not have understood the words, yet they understood everything. Kuniko returned to her seat and immediately turned to her oxygen.

The last hibakusha to testify was Kanji Kuramoto, president of the Committee of Atomic Bomb Survivors in the U.S. He took the witness stand to describe the discrimination survivors have suffered in every aspect of their lives, from insurance policies and employment to marriage. "The issue," Kuramoto concluded, "is whether the American government can assist a small number of the American survivors living today." Congressman Roybal responded to this testimony, "It is sad. It is moving, and I hope that it is read throughout the nation."

Following Kuramoto's testimony, Dr. Mitsuo Inouye, president of the Southern California Japanese American Medical Association, took the witness stand. Before reading his prepared text, which had already been distributed, he began, "I am here today, probably paying my dues to the atomic bomb survivors. When the atomic bomb fell . . . I was in the military intelligence ready to be assigned . . . to the invasion of Japan. I did heave a sigh of relief, which eventually gave way to guilt, and I hope I will pay my dues today." Inouye had been drafted into the U.S. Army from the Heart Mountain relocation center. His two elder brothers were already serving in Europe. When the war broke out, his parents had been anxious to go back to Japan, but their children had stopped them. If they had gone back, Inouye might very well have been one of the hibakusha. By the same token, any one of the U.S. hibakusha could have been serving in the U.S. Army and escaped the bombing had fate not led them to Japan. At the conclusion of his testimony, Inouye quoted the famous lines of John Donne:

> No man is an island, entire of itself; every man is a piece of the continent, a part of the main; if a clod be washed away by the sea, Europe is the less, as well as if a manor of their friends or of thine own were; any man's death diminishes me, because I am involved in mankind; and therefore never send to know for whom the bell tolls; it tolls for thee.

The hearing continued with the testimony of several scientists and academics, including the social psychologist Joe Yamamoto of UCLA, who described the deep psychic scars of the hibakusha and explained the psychology of those who live in constant fear. John Auxier, head of the Health Physics Division of Oak Ridge National Laboratory and the man responsible for establishing the system of measuring radiation dosage, also testified, though few in the audience understood his technical presentation.

The penultimate witness was Max Mont who, as vice president of the Community Relations Conference of Southern California, represented ninety-five organizations serving the public interest in the huge megalopolis of Los Angeles. His testimony was full of good will, and he warmly embraced the hibakusha as the "forgotten Americans." Later, Kaz Suyeishi expressed her appreciation, again with tears in her eyes, for "the very kind testimony of a white man." His testimony was significant, because without support from the American public beyond the Japanese American community, the passage of the bill would be extremely difficult.

Congressman Mineta, who had been posing very effective and pertinent questions, asked Mont, "Do you think that there would be concern about whether or not there would be people coming here from Japan just to take advantage of this bill after it passes?" Mont replied, "Japan provides much better, much more comprehensive, and much more continuing maintenance care for those who suffered the atomic blast and fallout, and it is very unlikely that any individual would come here for that reason." Of course, Congressman Mineta was well aware of this. He had asked the question, because he knew that this sort of question would be raised by his fellow members of Congress, and he wanted the answer in the record.

The final witness was Karl Nobuyuki, the national executive director of the Japanese American Citizens League. Two years earlier, at its national convention in Sacramento, the JACL had endorsed the bill but little support was forthcoming. In fact, when the Los Angeles chapter of JACL requested $300 to publicize the congressional hearing, the national headquarters turned it down. Nevertheless, Nobuyuki spoke strongly on behalf the bill for medical aid to the U.S. hibakusha.

When the hearing was adjourned it was already past one o'clock. Together with the three Congressmen, the witnesses went to lunch. Outside the building, the sky was perfectly clear and the bright California sun was shining.

Notes

1. This chapter is based on the author's personal observation of the hearings in Los Angeles as well as the official report, *Payments to Individuals Suffering from Effects of Atomic Bomb Explosions: Hearings Before the Subcommittee on Administrative Law and Governmental Relations of the Committee on the Judiciary*, House of Representatives, 95th Congress, Second Session on H.R. 8440, March 31 and June 8, 1978 (Washington: Government Printing Office, Serial No. 43, 1978).

2. Telephone interview with Karl Z. Morgan, 1976, Atlanta.

seventeen

"We Are All Hibakusha"

The hearing in Los Angeles represented success in the long effort of the American hibakusha to gain the attention of the Congress of the United States. William P. Shattuck, who had been on the platform as legal counsel to the subcommittee, did not miss a single witness. "What particularly impressed me," he later remarked, "was the human aspect, the fact that some American citizens are victims of the atomic bombs. A full analysis may belong in the realm of historians or biographers, but I was impressed by the extent of the disaster brought about to some individuals by a slight twist of history. Probably in all American history, no war since the Civil War has separated so many families so greatly. This fact became public for the first time through the testimony at the hearing."[1]

When news of the hearing reached Hawaii, atomic bomb survivors there began to contact Kanji Kuramoto, breaking the silence they had maintained up to that point. This broadened the base of support for the bill, but considerable lobbying would be necessary if the bill was to be passed by Congress. The hibakusha in America remained too weak to exert influence by themselves, but the JACL promised to start lobbying for the bill through its representative in Washington. In American society, acronyms are given to events or organizations that have become accepted as a social phenomenon, and it was a sign of progress that the JACL began to refer to the bill as the "ABS Bill," for the "Atomic Bomb Survivors Bill."

Even though Karl Nobuyuki, executive director of the JACL, described the bill as "strictly a matter for Japanese Americans,"[2] it would require broad support from peace groups and religious organizations to win passage. The ABS bill was intended primarily to help hibakusha with their medical costs, but it had a deeper and more far-reaching significance, for it implied that the U.S. government should accept responsibility for its ill-fated citizens who suffered from the atomic bomb in an enemy country during the war.

The assertion of this responsibility implicitly tied the hibakusha to a larger class of victims of the American nuclear program, some of whom were beginning to demand compensation during the late seventies.

The U.S. government and the majority of Americans rarely questioned the accepted wisdom that the atomic bombs dropped on Hiroshima and Nagasaki brought an end to the Pacific War and saved the lives of hundreds of thousands of Americans and Japanese. After the war, the United States was engaged without pause in developing nuclear weapons more destructive than those of the Soviet Union. By the end of the seventies, the worldwide nuclear stockpile amounted to twenty tons of TNT for every human being on the planet.

Beginning with a test detonation on Bikini Atoll in 1946, nuclear tests continued throughout these decades. When the U.S. government chose the Nevada desert as a nuclear test site, atomic bombs began to fall on the United States itself. However, several decades passed before this simple fact became evident to the public. Over the years, the United States has been producing a great many "hibakusha," and this fact was being discovered at last.

A large number of civilians and soldiers of the U.S. Army were present at nuclear tests in Nevada during the fifties. In January 1978, the House Commerce Subcommittee on Health and the Environment held a three-day hearing and heard the testimony of U.S. soldiers and scientists who participated in nuclear tests some twenty years before.[3] From 1946 to 1964, a total of 183 nuclear tests were carried out in the atmosphere by the U.S. government. Up to and including 1957, civilians and infantry units participated in eighteen nuclear detonations.

In the largest nuclear test, Operation Smoky, conducted in August 1957, a forty-four kiloton atomic bomb (more than three and a half times as powerful as the Hiroshima bomb) was detonated at Yucca Flats, a test site in Nevada. Soldiers observed the detonation from only a few miles from ground zero. Paul R. Cooper, one of the Smoky participants, recalled: "When I covered my face with my hands, I could see the bones in my hands like an X-ray at the moment of explosion."[4] Immediately afterward, thousands of soldiers were ordered to march over the hot ground to within one hundred yards of ground zero. According to the explanation given by the Defense Department, Operation Smoky was a practice exercise, a "war against fear—fear of the unknown—to prepare soldiers who might fight on an atomic battlefield." Leaving aside the question of whether this type of battle is conceivable in any future war, what the Defense Department was trying to obtain was "invaluable first-hand knowledge" of the physical and psychological effects of a nuclear attack on fighting units.

A total of 450,000 soldiers, scientists, and civilians participated in nuclear tests. Many of them were not even provided with radiation-film badges to measure radiation exposure. Very few follow-up surveys were ever made on the health of those who participated. Neither the Defense Department nor the AEC kept complete records of the participants in the tests. To make matters worse, the personnel records of these GI's were destroyed by a fire in a military warehouse in St. Louis, so that those who later wished to make claims were unable to establish their presence at the tests.

At that time, low-level radiation was not considered hazardous, despite years of research by the ABCC on radiation effects. Participants were led to believe there was a safe level of exposure to radiation. Since it can take upwards of twenty years before the effects of radiation begin to reveal themselves in the human body, the military and the AEC had plenty of time to pursue their programs for "peaceful uses of atomic energy" on the one hand and preparations for nuclear war on the other, exposing innocent soldiers and civilians to incalculable dangers without fear of embarrassing questions.

Twenty years after Operation Smoky in Yucca Flats, however, some of the participating GI's started to develop leukemia. Paul Cooper was one of them. He filed a claim for compensation with the Veterans Administration, contending that his disease was a direct result of Smoky. Dr. Glyn G. Caldwell of the U.S. Center for Disease Control (CDC) in Atlanta, which maintains oversight of disease and epidemics in the U.S. military, tracked down 450 of the three thousand Smoky participants and discovered that seven individuals in addition to Cooper suffered from leukemia. Statistically, only two cases of leukemia would be expected in a group of three thousand people exposed at the age of twenty-two, the average age of those participating in the 1957 test. Eight cases in a group of 450, therefore, indicated an abnormally high rate of incidence. Were not those soldiers from the Nevada tests also hibakusha?

Congressional hearings were subsequently held to review the effects of radiation on human health and to determine whether existing regulations, which stipulate "acceptable levels of exposure," were adequate to protect health. After 1956, annual exposure levels of one-half rem for the population at large and five rems for people who work with radiation had been considered safe by nuclear authorities. The Defense Department insisted that the average exposure of the participants in Smoky was only 1.25 rems, and that Cooper's exposure badge was in this range.

Karl Z. Morgan, former chief of the Health Physics Division at the Oak Ridge National Laboratory, testified that no dose of radiation could be low enough to assure that its risk of causing a malignancy was zero. He argued that each radiation dose, small as it may be, builds up in the human body and increases the risk of chromosome damage or cancer. He maintained that even a dose considered "safe" could trigger cancer and leukemia.

On August 3, 1978, the U.S. Veterans Administration announced that it would pay compensation to Donald Coe, a veteran who claimed that he developed leukemia because of the nuclear test in Nevada. The amount of compensation was reported to be between seven hundred and one thousand dollars a month. Paul Cooper, who had filed the first claim, was also ruled eligible for compensation, but he had died immediately after the hearings.

Radiation produced by the bomb tests was spread by the wind from the Nevada test site and into the neighboring states of Utah and Arizona. Not a few scientists warned of possible genetic or fatal effects of radiation on embryos and newly born babies, whose cells are extremely sensitive.

Around the time of the Operation Smoky hearings, alarming results of research on the dangers of low-level radiation were also made public. A study discovered a very high incidence of cancer among workers at the Hanford Nuclear Reservation, a massive nuclear facility in Richland, Washington. Hanford was originally built to produce plutonium for the atomic device tested in the desert of Alamogordo and for the atomic bomb dropped on Nagasaki.

Thomas F. Mancuso, a University of Pittsburgh professor, had been contracted by the AEC for a fourteen-year follow-up study on the health conditions of 35,000 workers in Hanford. His research showed that the radiation dosage that had been considered safe could cause cancer in certain parts of the body, including the bone marrow, lungs, and pancreas. This finding that a level of radiation far lower than the official "safe" level could cause cancer surprised the Department of Energy (DOE, the cabinet-level successor to the AEC and the ERDA). Mancuso's research was abruptly terminated, under pressure from the DOE, which also opposed the publication of his results. However, the discovery of these new hibakusha was widely reported in 1978 and had already become an accepted fact.[5]

On February 19, 1978, the *Boston Globe* reported that an alarmingly high rate of deaths from cancer had been observed among workers at the Portsmouth (New Hampshire) Naval Shipyard, a federal nuclear submarine plant.[6] Dr. Thomas Najarian of Boston's Veterans Administration

Hospital was the first to become interested in a former worker at the shipyard's nuclear facility who was hospitalized with leukemia. Hibakusha might also be found here, he thought.

Navy officials denied him access to lists of former workers, so Najarian, a hematologist, went through death certificates with the help of the *Boston Globe*. They found that former Portsmouth employees were contracting cancer at a rate four times higher and leukemia three times higher than industrial workers at large. Admiral Hyman G. Rickover, the "father of the nuclear submarine" and a mentor of then President Carter, disputed the research, but Congress urged the CDC in Atlanta to investigate.

On April 22, 1978, NBC-TV news reported that many former workers at the nuclear facility in Rocky Flats, Colorado, were suffering from cancer. Similar stories followed. It was increasingly evident that the nuclear industry in the United States had been producing new hibakusha in large numbers.

The International Symposium on Damage and Aftereffects of Atomic Bombing of Hiroshima and Nagasaki, organized by the international non-governmental organizations of the U.N. and meeting in Hiroshima in the summer of 1977, adopted the slogan "We are all hibakusha." Arthur Booth, chairman of the international preparations committee, declared in his opening speech at the symposium:

> We are all survivors of the Hiroshima and Nagasaki bombs. . . . Although we did not experience the blast and the burns, we all carry in our bodies man-made radioactivity which would never have been there but for the nuclear explosions which have followed since 1945. In more senses than we might care to believe, we are all hibakusha now.[7]

There were no delegates representing the American hibakusha at the symposium in Hiroshima. Not only did they have little ability as an organization to send delegates to Japan, they had little desire to be involved with the event. So great was the gap in consciousness between the Japanese and American hibakusha that they chose not to take the opportunity to appeal the plight of the American bomb survivors to the world. Fundamentally, the Japanese hibakusha were charging the United States with responsibility for dropping the atomic bombs; the American hibakusha were pursuing a more limited agenda and were unwilling to take the political risks of associating with the larger movement.

The size of this gap was illuminated in May 1978 when the American atomic bomb survivors rejected a request to meet in Los Angeles with a non-governmental delegation of Japanese on their way back from the

U.N. Special Session on Disarmament. Judging from the accounts of those involved, each side lacked understanding of the other. The Alliance for Survival, a peace organization that was hosting the Japanese delegation, sent a last-minute notice to representatives of the American survivors "to gather some fifteen hibakusha for tomorrow's meeting."[8]

The American hibakusha apparently saw this hurried request as egotism on the part of the Japanese delegation. They were greatly offended, and some concluded, "They (the Japanese delegation) are thinking only about their own convenience. They are only trying to use us. We are not guinea pigs. While the Japanese bomb survivors are fortunate enough to be able to form a delegation of five hundred people and come over to the U.S., we cannot even afford to go to Japan to receive medical treatment."[9] The American atomic bomb survivors were also disappointed because the Japanese delegation, which claimed to value unity and mutual understanding, made no mention of them at the United Nations forum.

Also behind the refusal to meet with the Japanese delegation was the American hibakusha's fear of, or distaste for, political involvement. A hibakusha activist said:

> They are foreigners and we are Americans, even though we have the same Japanese face. In addition, our country dropped the atomic bombs. They come here and make their protest, and then go back to Japan. Not all Americans take offense at their activities, but some do. And we are the ones who have to deal with the backlash. We are not demanding that the U.S. government take responsibility for dropping the bombs, and we're not demanding compensation either. We are simply asking the government to pay the medical expenses for the atomic bomb disease which might strike us at any moment."[10]

There was little basis of unity to be found here. It might be easy to criticize the American hibakusha as lacking in social conciousness. But what the survivors were seeking was to be accepted as American citizens, not only in principle, but in practice, through passage of the medical assistance bill. In other words, they were asking the government to bridge the gulf that was created in their hearts by their separation from their home country during the war.

On March 30, 1978, a day before the Los Angeles hearing, the Japanese Supreme Court made a historic decision in the case of Son Gin Doo, a Korean atomic bomb survivor who had entered Japan illegally in order to receive medical treatment for his atomic bomb disease. He was granted the right to obtain a hibakusha ID card.[11] The Supreme Court ruled that the Medical Treatment for Atomic-Bomb Victims Act was based on a "humane purpose," and "that due to the peculiarity and seriousness of the suffering caused by the atomic bombing, the act does not discriminate against any survivor on the basis of nationality."

At the time of this court decision, even some American newspapers praised it in their editorials. The *Denver Post*, for instance, wrote: "The Supreme Court of Japan has ruled that all persons injured by the A-bombs, regardless of nationality, are entitled to medical care. Can the United States do less for its own citizens who suffered the same misfortune?"[12]

The American hibakusha and their friends responded in much the same way. The Japanese government had accepted responsibility and guaranteed medical treatment to a non-Japanese hibakusha, even one who had been in the country illegally. Why, the hibakusha wondered, could the U.S. government not provide treatment for its own citizens who happened to be survivors of the atomic bomb?

With this issue still unresolved, it was difficult for the American hibakusha to develop a sense of unity with the Japanese hibakusha, or with the new "hibakusha" that were emerging on the national scene.

Mitsuo Inouye, president of the Japanese American Medical Association in Southern California, made a deep impression at the hearing in Los Angeles when he closed his testimony with the lines of John Donne: "And therefore never send to know for whom the bell tolls; It tolls for thee." He highlighted the fact that humankind shares a common destiny, especially when the threat of nuclear holocaust was hanging over the entire world.

Both the Japanese Supreme Court decision in the case of Son Gin Doo, and the International Symposium on Atomic Bomb Aftereffects with its slogan, "We are all hibakusha," represented bells tolling a message for all mankind. The American hibakusha had suffered physically from the atomic bombs dropped by their own country, and they had suffered the psychic pain of being told by their fellow Americans: "You were the enemy." Yet upon their shoulders rested a historical mission as well—to toll an alarm bell in their own country, jointly with all the hibakusha of the world, for the sake of the future of humankind.

Notes

1. Telephone interview with William Shattuck, April 10, 1978, Washington, D.C.
2. Interview with Karl Nobuyuki, April 6, 1978, San Francisco.
3. *Effects of Radiation on Human Health*, hearings before the Subcommittee on Health and the Environment of the Committee on Interstate and Foreign Commerce, U.S. House of Representatives, January 24–26, February 8, 9, 14, and 28, 1978.
4. *Newsweek*, February 6, 1978.
5. On Mancuso's account, see Howard L. Rosenberg, *Atomic Soldiers: American Victims of Nuclear Experiments* (Boston: Beacon Press, 1980), pp. 151–152.
6. *Boston Globe*, February 19, 1978.

7. On the Hiroshima symposium, see *Asahi Shimbun*, August 1, 1977.
8. Interview with Kaz Suyeishi, August 26, 1978, Los Angeles.
9. Conversation with Kuramoto, August 31, 1978, San Francisco.
10. Interview with Suyeishi.
11. *Asahi Shimbun*, March 31, 1978.
12. *Denver Post*, editorial, May 8, 1978.

eighteen

Epilogue: Fifty Years After the Bomb

An exhibition on the atomic bombs and the end of World War II was planned for the Smithsonian Institution's National Air and Space Museum in Washington in 1995, fifty years after the bombings. The exhibit was to be mounted alongside the restored fuselage of the *Enola Gay*, but a fierce controversy developed over its content. The draft plan for the exhibit was indeed a brave undertaking, posing a number of questions: Was President Truman's decision to drop the bomb justified? What were its human consequences? What alternatives, if any, existed for ending the war? And, did the bombs usher in the postwar nuclear arms race?

As might have been expected, veterans organizations and defense lobbies, led by the American Legion and the Air Force Association, objected vehemently to the plan. Working with conservative politicians, they applied powerful pressure on the museum staff to revise the content of the exhibit. On the other hand, historians and peace organizations criticized these political efforts to manipulate public history. In the end, the military won a pyrrhic victory. Plans to reveal the human consequences of the bomb and to explore alternatives to it were scrapped. When the exhibit finally opened in June 1995, all that remained was the front half of the *Enola Gay* and a brief video account by its crew members.[1]

The controversy provided a window on the gap between scholarly understanding of atomic issues and official narratives on the atomic bombing. The latter was expressed in a Senate resolution, passed unanimously on September 19, 1994, condemning the Smithsonian's original plans. "The role of the *Enola Gay* during World War II," this read, "was momentous in helping to bring World War II to a merciful end, which resulted in saving the lives of Americans and Japanese."[2] The reasoning behind the resolution was that World War II was a "good war" that America fought in response to Japanese aggression, and that the atomic bombing that ended the war was fully justified. This reasoning, long contested by historians of the bomb, is almost akin to an official religion in the United States.

After the board of the Smithsonian decided in January 1995 to scale back the exhibition, the national controversy continued. President Clinton rubbed salt into the wound for many Japanese and for all hibakusha by stating, "The United States does not need to apologize to Japan for dropping the bomb. President Truman made the right decision."[3]

To the survivors of Hiroshima and Nagasaki, the dropping of the bomb was far from being a "merciful" act. No matter how Truman's decision may be justified politically or strategically, the survivors could never agree that it was the "right" decision. To American survivors, moreover, these developments—the Senate once again expressing the righteousness of the atomic bombing, and the president as the head of the nation confirming it—reminded them of how hopeless their situation continued to be. Fifty years after the fact, there had been virtually no change in American official opinion on the atomic bombing as represented by Congress and the president, and supported for the most part by the press. Scholarly critics and writers have effectively challenged the official story from a variety of perspectives. Moreover, in influential writings such as John Hersey's *Hiroshima*, Americans have been exposed to victims' perspectives on the bomb. Whatever the scholarly or popular doubts, however, the official story remains the dominant narrative.[4]

Given this situation, it is hardly surprising that the effort over the years by American hibakusha to obtain modest medical aid from Congress amounted to trying to open a small hole in a hard, thick wall.

In the spring of 1978, when the congressional subcommittee hearing on the medical aid bill was held in Los Angeles, all the witnesses had been supportive and the mood was optimistic about the bill's passage by Congress. However, the House Judiciary Committee failed to recommend the bill, and it went nowhere. A similar bill was introduced in every Congressional session in the following years, but it died in committee every time. After Congressman Roybal, the initiator and constant sponsor of the bill, retired in 1992, no new champion emerged and the bill was not reintroduced.

What happened? To find an answer, we need to go back to the second hearing for the hibakusha bill held in Washington on June 8, 1978, several months after the first hearing.[5] There were only two witnesses, Congressman Norman Mineta, a cosponsor of the bill, and Harry Takagi, the JACL Washington representative. According to Congressman George Danielson, chairman of the subcommittee in charge, both the Pentagon and the Department of Health, Education and Welfare were asked to send representatives, but they declined.

At this hearing, one committee member raised an issue that may have sealed the fate of the bill. Looking back, one can infer that his comments reflected the positions of the two departments that declined to send representatives. The committee member was Representative Carlos Moorhead, a conservative Republican and World War II veteran from Glendale, California.

"You know," Moorhead stated, "we do not award the victims of war under any circumstances that I know of at the present time, and the Federal Tort Claims Act, the Military Claims Act, precludes liability on the part of the United States for injuries. So that is the decision that has to be made, whether this should be made a special case."

The reader may recall here the letter sent to Kanji Kuramoto, president of CABSUS, from the head of the State Department's Japan desk, which read in part: "It has been the long-standing policy of the United States government, however, not to pay claims, even on an ex-gratia basis, arising out of the lawful conduct of military activities by U.S. forces in wartime."[6]

It is only since the end of World War II, by the enactment of the Federal Tort Claims Act, that citizens have been able to ask the federal government to pay compensation for damage inflicted by faults or illegal acts on the part of the government or its employees.[7] This law has many exceptions, however, including military conduct in wartime and incidents that take place outside of the country. Since the government regards the atomic bombings as lawful military conduct, the Tort Claims Act cannot be applied to Hiroshima and Nagasaki. What American hibakusha had been gingerly asking the government for was medical assistance on an ex gratia basis, but the letter from the State Department confirmed that the U.S. government does not "pay claims, even on an ex gratia basis." Congressman Moorhead pointed out that giving medical aid to American survivors would create an exception to this long-standing policy.

Congressman Mineta responded that the postwar German government has been paying medical expenses and living assistance to Jews who were persecuted by Nazi Germany. Mineta further asserted that the atomic bomb survivors were the first victims of nuclear war, yet the American survivors could not receive any compensation from their own government. They would have to visit Japan, he said, in order to obtain medical treatment, or pay exorbitant medical bills out of their own pockets.

Unconvinced, Congressman Moorhead posed another question: "We had thousands of people, of our citizens, trapped in Germany and in Italy that were hurt in the big major bombings of Hamburg and other cities there, who have never regained their health. Should they be included also?"

"I think there is a difference," Mineta responded, "of whether or not there is the utilization of, let's say, the regular weapons of war versus a nuclear weapon, which has long-lasting genetic as well as other kinds of injuries. . . . With radiation, the effects are really long lasting, the extent unknown. The victims of atomic bombs have much greater medical expenses than those who might have injuries or permanent disabilities arising from conventional weapons."

Moorhead asked no more questions. It was as though he had said enough. At the beginning of his comments, he had indicated that HEW did not send witnesses because they were against the bill. However, according to a well-informed former JACL official, the bill's strongest opponent was the Pentagon. Right after the second hearing, according to this account, the Defense Department sent a short memorandum to those involved, indicating clear opposition to the bill. "That effectively stopped the whole move," the JACL staffer said.[8]

The Pentagon usually does not express an opinion in writing on a matter as delicate as the hibakusha bill, and I have not been able to locate a copy of this memorandum. But the military would never allow a precedent to be set whereby civilians could be compensated for damage inflicted in wartime. This would invite a flood of claims by citizens who had been hurt in previous wars, and its future implications for waging war, nuclear or otherwise, would be tremendous.

In any case, after the second hearing, the bill never reached the full Judiciary Committee, and it eventually died.

Another episode illustrated the government's inclination to distance itself from the medical needs of the American hibakusha. In the spring of 1979, when doctors from Hiroshima made their second visit to the United States to perform medical check-ups on hibakusha, the support group in San Francisco asked the Department of Health, Education and Welfare for permission to use the facilities at the Public Health Hospital, a federal institution. The Assistant Surgeon General's office turned down the request.

California Senator Alan Cranston intervened with a letter to the secretary of HEW asking him to reverse the decision. "I understand," Cranston's letter noted, "that the Office of Japanese Affairs of the Department of State was consulted for an advisory opinion and that the denial is due to a long-standing Executive Branch policy whereby the Federal government disallows utilization of Federal facilities because of possible Federal tort liability."⁹ Cranston's letter expressed concern that "the Federal government's posture here may be out of touch with recent developments in the complicated area of radiation exposure," and he specifically pointed to the report of a task force that HEW itself had put together to study the spreading problem of radiation exposure. The task force focused on soldiers and civilians exposed to radiation from nuclear tests and workers in nuclear industries.

Cranston had succeeded in getting the task force to address the question of the hibakusha in the United States, although it limited its concern to the estimated five hundred to one thousand who were American citizens at the time of atomic bombing, excluding Japanese hibakusha who subsequently took up residence in the United States. A draft report Cranston sent to Kuramoto stated, "The work group does not have extensive information about this group, whose experiences should be investigated further. However, it may be appropriate to provide medical care to them, at least for diseases associated with radiation exposure."¹⁰

While only a small step, this was the first indication that the Carter administration might consider medical treatment for the hibakusha. HEW eventually gave permission to use the San Francisco hospital for the hibakusha exams, but the idea of providing or subsidizing medical treatment for hibakusha along with other radiation victims died with the Carter administration itself.

✍ ✍ ✍

During the early eighties, American survivors of Hiroshima and Nagasaki were included in many of the efforts to focus national attention on the problems of radiation exposure and nuclear war.

In April 1980, the National Citizens' Hearings on Radiation Victims were held in Washington for four days.¹¹ The event was organized by a coalition of peace groups and involved prominent scientists, religious leaders, and peace activists, including Nobel laureates Linus Pauling and George Wald. Witnesses included "atomic soldiers" and civilians from nuclear test sites in Nevada and the Pacific Islands; workers from uranium enriching facilities, nuclear power plants, and nuclear submarine dockyards; and Japanese American survivors of Hiroshima and Nagasaki, including Kanji Kuramoto of CABSUS.

The report issued after the hearings estimated that America's radiation victims numbered more than one million citizens. After President Ronald Reagan took office, however, there was little hope of government action to address the problem of radiation exposure. Most radiation victims had to mount involved court battles in order to win compensation, with their claims based on the negligence of the government or private contractors.

Successful cases were fought by atomic soldiers and even by Bikini islanders. This avenue, however, was not available to the hibakusha, because their exposure to radiation occurred during wartime.

Kuramoto and Kaz Suyeishi, as representatives of the American hibakusha, also participated in the Non-Governmental Organizations (NGO) meetings, held in New York City in June 1982 on the occasion of the second United Nations Special Session for Disarmament. A documentary film on the American hibakusha was shown at one of the meetings. Entitled "Survivors" and directed by Steven Okazaki, a San Francisco Sansei, the film was completed in April of that year and aired nationally on PBS the following August.

Okazaki and other supporters had begun concerted efforts to bring the hibakusha's story to a wider public audience. They helped CABSUS produce a 1979 pamphlet, "America's Bomb Survivors: A Plea for Medical Assistance," which was distributed to every JACL chapter in the nation. In January 1982, a group of supporters centered in San Francisco organized themselves into Friends of Hibakusha (FOH). Rev. Nobuaki Hanaoka, a minister at a San Francisco Methodist church, was elected president of FOH.

FOH began publishing a newsletter, *The Paper Crane*, in January 1983 to publicize the problem of the survivors. Members also started compiling oral histories of individual survivors.[12]

The most notable contribution of FOH was to create links between American survivors and the new radiation victims. By spring 1985, *The Paper Crane* reported, a total of 1,125 lawsuits had been filed against the federal government by these radiation victims, seeking $2.5 billion in compensation. FOH believed that survivors of Hiroshima and Nagasaki should work in concert with these nuclear activists, and together with the National Association of Radiation Survivors (NARS), FOH sponsored a Radiation Survivors Congress in San Francisco December 12 to 14, 1984.[13]

In addition to the victims of radiation exposure from U.S. nuclear tests and the nuclear industry, Katsuyoshi Komatsu of Hiroshima and Ryuzo Fukabori of Nagasaki came from Japan, and Lee Silgun attended as a representative of the Hiroshima survivors in Korea. Partly because of Lee's presence, the congress was made aware that some Korean atomic bomb survivors were now living in the United States. Koreans were immigrating to the United States in large numbers, so it was to be expected that some of the immigrants were survivors of Hiroshima, where an estimated twenty thousand Koreans were exposed to the bomb.[14]

One result of the congress was that CABSUS was able to hold a national meeting for the first time in its history. CABSUS was continuing to have difficulty organizing itself as a coherent national organization, in part because of the geographical separation of its membership. In California, the group had polarized between the north (the San Francisco area) and south (the Los Angeles area), as is common among political organizations in the state.

CABSUS also continued to struggle over the political direction of its activities. Although the new friends that the members made at the Radiation Survivors Congress were suing the American government and organizing demonstrations to change government policies on nuclear matters, CABSUS found it difficult to join them in either activity. Thus, despite the efforts of Reverend Hanaoka and others in FOH, the link with later victims of radiation made at the Radiation Survivors Congress of 1984 did not last.

The bias of survivors against involvement in political campaigns was stronger in southern California. In the summer of 1984, when Los Angeles hosted the Olympic Games, national peace and justice groups organized a two-week counter event called "Survivalfest," which culminated in a gathering of some twenty thousand participants in MacArthur Park on August 5. A "peace flame" was brought over from Hiroshima's Peace Park by a survivor and used to light an eternal flame. Shigeko Sasamori, one of the "Hiroshima Maidens," made a passionate appeal for the abolition of all nuclear weapons. But survivors living in the Los Angeles area had no visible presence in the events.

There was an independent southern California counterpart to FOH: the Friends of Atomic Bomb Survivors (FABS). Among other activities, it hosted a stage play, "Visitors From Nagasaki," written by Perry Miyake, Jr., at East West Players in Los Angeles in March 1984. The proceeds were contributed to the Hibakusha Travel Fund, to support survivors' trips to Japan for medical care.

However, CABSUS activities in southern California were limited, for the most part, to the efforts of Kaz Suyeishi, including tireless lecturing, TV appearances, and technical consultations for movie productions on Hiroshima. As valuable and moving as her efforts were, they did little to strengthen the organization of the survivors.

The Japanese American Citizens League had placed support for the hibakusha on its official agenda in 1976, but the effort was always a low priority. At the top of JACL's agenda was the redress campaign, the campaign to obtain compensation for damages inflicted by the evacuation of Japanese Americans from the West Coast to relocation camps during World War II. This campaign came to fruition in 1989 when Congress enacted a law requiring payment of twenty thousand dollars to each survivor of the internment camps.

The redress campaign had the solid support of Japanese Americans nationwide. It was led by Hawaii Senators Daniel Inouye and Spark Matsunaga, and by California Congressmen Norman Mineta and Robert Matsui. Its success was assured when the campaign secured the support of the Texas congressional delegation. This was a form of thanks to the all-Nisei battalion of the 442nd Regiment that rescued a trapped Texas regiment at great sacrifice during World War II.

President Ford had made a formal apology in 1975, declaring the relocation a mistake. The Supreme Court, which had declared Executive Order 9066 a necessary wartime measure, reopened the case after the war and found the executive order to be a violation of fundamental human rights, hence unconstitutional. Compensation was seen as restoring "justice denied" and was widely supported by public opinion.[15]

However, the dropping of the atomic bomb was a totally different issue. The survivors were Nisei who by chance found themselves on the wrong side in the war. Furthermore, the U.S. government continued to claim that it was right to drop the bombs, and there was no sign that it would ever revise this position. Congressman Mineta had warned that it would be difficult to win passage of the bill, especially if the Pentagon was determined to block its passage. The survivors organization had no lobbyist, few members, and limited financial resources. Looking back, the fate of the bill to aid the American hibakusha had been sealed from the beginning. As Kuramoto remembers it, Congressman Mineta told him, "The last resort is the Japanese government."

~ ~ ~

Under a program that began in 1980, the Japanese government extended assistance to Korean hibakusha who traveled to Japan to obtain medical care. The Korean government provided travel expenses, and the Japanese government assumed the financial burden, medical expenses included, for two months of hospitalization. A total of 349 Korean hibakusha benefited from this program before it expired in November 1986.

After the program ended, Seoul increased domestic efforts to provide care for its survivors by building more clinics and expanding financial assistance to pay medical bills. Tokyo provided more than four billion yen (about $25 million) between 1989 and 1993 to support this effort. Although the Japanese government was not willing to send doctors to Korea, had it not been for Japan's colonial rule of the Korean peninsula, Koreans would not have been in wartime Japan and there would have been no Korean victims of the atomic bombing.

Mistakenly believing that the Japanese government was also paying the travel costs of the Korean hibakusha, Kuramoto appealed to Prime Minister Yasuhiro Nakasone to provide travel assistance to survivors living in America.[16]

In fact, financial aid to American survivors who came to Japan for medical care had started on local government and private levels, though on a

Epilogue: Fifty Years After the Bomb

small scale. In July 1982, the city of Nagasaki began inviting two survivors a year to visit at public expense. In Hiroshima, a prefectural assemblyman, Akira Ishida, himself a survivor, learned about the plight of American survivors while attending the second United Nations special session in June 1982. Upon his return, Ishida established a fund to support the "homecoming" of hibakusha for medical treatment. Thirteen survivors benefited from this fund over the next five years. After 1988 this program was taken over by the Hiroshima Prefectural Medical Association. Resources were provided to bring five survivors a year to Hiroshima for treatment for the next ten years.[17]

The trips to Japan for medical care were available to only a small number of hibakusha. Most of the American survivors benefited far more from the visits of doctors from Japan to provide routine medical checkups, supported by funds from the Japanese Ministry of Health and Welfare and the volunteer services of members of the Hiroshima Prefectural Medical Association.

After the first visit in March and April of 1977, the team of doctors came every other year to the cities where survivors were concentrated: Los Angeles, San Francisco, and Seattle. As in the case of the second visit, when HEW initially refused access to the federal hospital in San Francisco, there was often some trouble with U.S. authorities over the issuance of visas or the unscheduled checking of doctors' licenses. But the visits continued without interruption. In order to receive the doctors from Japan, support groups were organized among American doctors and citizens to handle the paper work, correspondence, and auxiliary help during the checkups. Funds were also raised to defray travel expenses for hibakusha coming from other states, and free housing arrangements were made.

The first medical checkups covered 123 survivors, and the second visit covered 150. The third visit in 1981, covering 203 survivors, extended its purview to Hawaii in the name of "health counseling." A Hawaiian chapter of CABSUS had been established the previous year, and thereafter medical checkups in Honolulu became routine. The fourth visit in 1983 covered 305 survivors, including survivors living in Canada who came across the border to Seattle. The Canadian government issued a special license to the visiting doctors on their fifth visit in 1985, and they were able to conduct exams in Vancouver. A total of 339 survivors were examined on this trip. The island of Maui in Hawaii was added for the sixth visit in 1987, and a total of 379 survivors received checkups. The seventh visit (1989) covered 406 survivors, the eighth (1991) saw 532, and the ninth

(1993) covered 549. In June 1995, the Japanese doctors marked their tenth visit, conducting 463 examinations. "Three visits will be maximum," one of the doctors from Hiroshima had confided to Kanji Kuramoto during the first visit. "That's all we can manage." It was a pleasant surprise that the visits continued.

There are a number of reasons that the number of survivors receiving checkups increased steadily over the years. The children of survivors began to participate in the exams, seven in 1989 and forty-nine in 1991. In addition, there was growing concern among the hibakusha over possible late effects of radiation, and individuals who initially had not wanted to be identified as survivors emerged as they sensed the desirability of a checkup. In Hawaii, a Nisei veteran who had participated in nuclear tests in the Pacific showed up and received a checkup.[18]

The Japanese doctors provided psychological relief along with the physical examinations of the survivors. In the rare instances where bomb-related symptoms were discovered, survivors could count on being sent to Japan for further treatment, with assistance from the cities of Hiroshima and Nagasaki. This was a source of great comfort, but it also represented a problem for the campaign seeking to secure medical assistance from the U.S. government. As long as the hibakusha were assured of periodic, free medical checkups by Japanese specialists, there was little motivation to press for U.S. government support.

Membership in the survivors group was never a prerequisite for participating in the biennial examinations by the Japanese doctors, but many of the hibakusha kept their membership current because they believed it guaranteed the regular checkups. In the summer of 1992, CABSUS split and the two largest chapters, Los Angeles and Hawaii, formed a new organization, the American Survivors Association (ASA), in March of the following year. Keichu Teramoto was elected the first president.

Despite the organizational split, the Japanese doctors continue to come to perform regular checkups, and all survivors can take advantage of this service. In this sense, as one knowledgeable doctor put it, "American survivors can afford to lose their nationwide organization."[19] Neither group holds out any further hope of assistance from the American government.

The saga of the American survivors of Hiroshima had reached a closure of sorts as the fiftieth anniversaries of Hiroshima and Nagasaki came and went. The survivors' most pressing need—medical attention—was being addressed, albeit without the U.S. government assistance and institutional support envisioned in the campaign for the medical aid bill. The desire to be acknowledged as American victims of the bombings had been satisfied to a degree in the course of the campaign, with the official support of such organizations as the JACL and the AMA, a moderate amount of media attention, and the activities of FOH and other organizations. To be sure, the existence of the American hibakusha had barely penetrated the national consciousness, but it could no longer be said that the survivors of were entirely "forgotten Americans."

Another factor that was working toward a sense of closure among the survivors in America was age: the median age of those who held U.S. citizenship at the time of the bombing was approaching sixty-five years. This meant that their numbers were declining and those who survived would soon become eligible for Medicare, if they were not already, and they would have fewer worries about paying medical bills.

The past two decades had also seen a dramatic shift in the political environment. The threat of nuclear war and the dangers of nuclear power were issues of great public concern in the late seventies and early eighties, culminating in the historic demonstration of one million people in New York on the occasion of the second United Nations Special Session for Disarmament in the summer of 1982.[20] The nuclear freeze movement and European campaigns against the deployment of nuclear weapons contributed to a ground swell of opposition to the nuclear establishment. The disaster at the Chernobyl nuclear power plant in the Soviet Union in April 1986 increased doubts about the safety of nuclear energy, which had already been thrown into question by the accident at the Three Mile Island reactor in 1979. The tide of history had shifted against the nuclear industry.

In December 1987, President Reagan and Soviet leader Mikhail Gorbachev signed a treaty to abolish intermediate-range nuclear weapons, followed in May 1988 by Reagan's visit to Moscow and the start of negotiations to reduce strategic nuclear weapons. In 1989, the year George Bush was inaugurated president, an avalanche occurred in world politics. The Berlin Wall was breached in November, and the Soviet Union collapsed soon afterwards, bringing an end to the Cold War. The danger of nuclear confrontation subsided, at least in the public consciousness. The threat of nuclear proliferation by smaller countries remained, but the world was no longer faced with two superpowers, confronting each other and armed to the teeth. The peace movement in the United States rapidly declined, and public concern shifted to environmental issues, poverty, and race relations.

As nuclear issues dropped off the public agenda, the possibility that the American survivors' medical needs would be addressed shrank even further. On the other hand, arms reduction relieved the concern of many hibakusha that the tragedies of Hiroshima and Nagasaki would be repeated. As Reverend Hanaoka, president of FOH, put it, "With the Cold

War over and the Soviet Union dissolved, the antinuclear movement in America lost its momentum. But I gather that atomic-bomb survivors feel that their efforts had some effect. I tend to agree that, in the larger context, the survivors' goals have been fulfilled."[21] While this comment struck me initially as a valiant effort to put a positive spin on the situation, many of the survivors I spoke with shared Hanaoka's view.

Will these survivors now merely fade from the public consciousness, now that the fiftieth anniversary of the bombings of Hiroshima and Nagasaki has passed? I hope this does not happen. The Smithsonian controversy reminded us that the national debate over the bombings remains unresolved, and how critical it is to maintain awareness of the reality of the human consequences of Hiroshima and Nagasaki. As long as nuclear weapons exist on this earth, the symbolic importance of the atomic-bomb survivors will remain strong. Their presence in the United States, as the original witnesses to the reality of the Nuclear Age, must not be forgotten.

Notes

1. For discussion of the controversy, see Edward Linenthal and Tom Engelhardt, eds., *History Wars: The Enola Gay and Other Battles for the American Past* (New York: Metropolitan, 1996). *The Journal of American History* 82:3 (December 1995) devoted the entire issue to the topic "History and the Public: What Can We Handle?"

2. The text of Senate Resolution 257 is in the *Congressional Record*, September 19, S. 12968, and can also be found in *Bulletin of Concerned Asian Scholars* 27:2 (April–June 1995). Most of this issue of *BCAS* is devoted to the Smithsonian controversy: "Remembering the Bomb: The Fiftieth Anniversary in the United States and Japan."

3. *New York Times*, April 8, 1995.

4. For a fuller discussion of these issues, see Laura Hein and Mark Selden, eds., *Living with the Bomb: American and Japanese Cultural Conflicts in the Nuclear Age* (Armonk: M. E. Sharpe, 1996); and Michael Hogan, ed., *Hiroshima in History and Memory* (Cambridge: Cambridge University Press, 1996).

5. *Payments to Individuals Suffering from Effects of Atomic Bomb Explosions: Hearings Before the Subcommittee on Administrative Law and Governmental Relations of the Committee on the Judiciary*, House of Representatives, 95th Congress, Second Session on HR 8440, March 31 and June 8, 1978 (Washington: Government Printing Office, Serial No. 43, 1978).

6. Letter from the State Department to Kuramoto.

7. The Federal Tort Claims Act was enacted in 1946 as Title IV of the Legislative Reorganization Act.

8. Telephone conversation with a former JACL official in California, April 5, 1995.

9. Senator Cranston's letter is in Kanji Kuramoto's personal papers (hereafter cited as "the Kuramoto files").

10. From the Kuramoto files.

11. A summary report of this hearing was carried in the May 1980 issue of *SANE World*, a Newsletter of Action on Disarmament and the Peace Race.

12. Recorded interviews and other related materials are archived at the Bancroft Library of the University of California, Berkeley. Accounts of the oral history project were given to the author by Dorothy Stroup in a telephone conversation, March 27, 1995, San Francisco. See also Kenzaburo Oe, *Hiroshima no "Inochi no ki"* (Hiroshima's "tree of life") (Tokyo: Iwanami Shoten, 1991), p. 162.

13. Telephone interview with Nobuaki Hanaoka, March 27, 1995, San Francisco.

14. The estimate of the number of Korean hibakusha is given in Sadao Kamata, ed., *Hibaku Chōsen-Kankokujin hibakusha no shōgen* (Testimonies of Korean Survivors) (Tokyo: Asahi Shimbunsha, 1982), p. 14.

15. On the redress campaign, see Yasuko Takezawa, *Breaking the Silence: Redress and Japanese American Ethnicity* (Ithaca: Cornell University Press, 1996); and Leslie T. Hatamiya, *Righting a Wrong: Japanese Americans and the Passage of the Civil Liberties Act of 1988* (Stanford: Stanford University Press, 1993).

16. Copy of Kuramoto's letter to Nakasone is in the Kuramoto files.

17. "Chronology of the Committee of Atomic Bomb Survivors in the U.S." (draft in Japanese) in the Kuramoto files.

18. The summary of Japanese doctors' visits is based on "Chronology."

19. Interview with Kenji Irie, April 1, 1995, Los Angeles.

20. Rinjiro Sodei, "Hankaku ni moeta New York" (New York City in the Heat of the Anti-Nuke Movement), *Hankaku no Amerika* (Anti-Nuke America) (Tokyo: Ushio Shuppansa, 1982).

21. Telephone interview with Hanaoka.